Land Restoration: A Rational, Grounded and Pragmatic Approach

V. Santhakumar · Ritika Jain ·
Muralee Thummarukudy

Land Restoration: A Rational, Grounded and Pragmatic Approach

Springer

V. Santhakumar
Azim Premji University
Trivandrum, Kerala, India

Ritika Jain
Centre for Development Studies
Trivandrum, Kerala, India

Muralee Thummarukudy
United Nations Convention to Combat Desertification (UNCCD)
Bonn, Germany

ISBN 978-3-032-26651-4 ISBN 978-3-032-26652-1 (eBook)
https://doi.org/10.1007/978-3-032-26652-1

This Springer imprint is published by the registered company Springer Nature Switzerland AG
The registered company address is: Gewerbestrasse 11, 6330 Cham, Switzerland

Contents

Part I
Introduction and Context

Chapter 1
Land Restoration: What Is It? Why Is It Important?

Introduction

Land is the most important natural resource for human and other beings. It is the top layer of the earth's crust formed by mineral particles, organic matter, water, air and living organisms (European Commission, 2006). It is the earth's terrestrial surface, and includes all attributes of the biosphere immediately above or below this surface. It enables human life. It provides the physical space for almost all human activities. It supplies not only food and fibre to human and other beings but also essential ecosystem services such as regulation of biochemical, biophysical and water cycles and nutrient filtering.

Land degradation has been taking place in different parts of the world and its negative impacts are acknowledged widely. One estimate notes that upto 40 per cent of global land area is degraded (UNCCD, 2024).[1] It is estimated that more than 2 billion hectares of land have undergone degradation.[2] The negative impacts of land degradation on nature and society are well known (Barrow, 1991; Sharafatmandrad & Mashizi, 2021). These include deforestation, the loss of biodiversity and other ecosystem services including the reduced ability to absorb carbon dioxide.[3] The major proximate causes of land degradation include deforestation, wetland drainage, overgrazing, and the expansion of agriculture and industrialisation and urbanisation (UNCCD, 2017).

The need to restore land so that it can serve its role in enabling the life of human and other species is recognised. 'Land restoration is the ecological process to restore a natural and safe landscape for humans, wildlife, and plant communities. This process paves the way to protect our ecosystems, create economic development, help prevent natural disasters such as floods, and increase soil productivity and food supplies. In

[1] https://news.un.org/en/story/2024/12/1157651.

[2] https://www.worldenvironmentday.global/did-you-know/key-messages.

[3] https://data.unccd.int/.

V. Santhakumar et al., *Land Restoration: A Rational, Grounded and Pragmatic Approach*, https://doi.org/10.1007/978-3-032-26652-1_1

short, land restoration is vital if we are to protect the environment, build resilience to drought, and help feed a growing global population'.[4] It 'is the process of ecological restoration of a site to a natural landscape and habitat, safe for humans, wildlife, and plant communities'.[5] Needless to mention its importance in protecting biodiversity and addressing climate change. Nearly 80 per cent of the carbon stored in terrestrial ecosystems is found in soils'.[6] The economic benefits of land restoration are huge. The United Nations Environment Program (UNEP) estimates that every dollar invested in restoration generates up to USD 30 in benefits.[7]

There are a number of initiatives to reduce or avoid land degradation. UNCCD has argued for Land Degradation Neutrality (LDN) by 2030. Bonn Challenge - a joint initiative by the German government and the International Union for Conservation of Nature (IUCN) started in 2011 with a target of restoring 150 million hectares of land by 2020, was then extended to 350 million hectares by 2030 (Verdone and Heidi, 2017). There are regional initiates too. For example, the one in Africa aims at restoring 100 million hectares by 2030. The Middle-East Green Initiative has a goal of restoring 240 million hectares. There are similar initiatives in Latin America, Europe, Caucasus, and Central Asia. The G20 Global Initiative on Land has the goal of reducing degraded land by 50 per cent by 2040.

However, these initiatives to reduce or mitigate land degradation face different challenges. First is the lack of adequate financial resources for ecosystem restoration compared to those activities which may have a negative impact on nature (UNEP, 2023). There can be contradictions between public policies and private practices leading to a higher level of degradation (UNCCD, 2022). The incentives and subsidies for agricultural development could be examples. Though private sector has the potential to carry out restoration activities, a major part of this is yet to be realised. The policies and actions to protect ecosystems (including land) are yet to be linked adequately to other development goals (UNEP, 2023).

Hence there is a need for a rational, grounded and pragmatic approach towards land restoration. It is to be part of the discourse of international and national policy-makers, government officials and private and other non-governmental actors who are concerned about the degradation of land. Detailing such an approach and making it part of a public discourse are the purposes of this book.

---

[4] https://www.unccd.int/land-and-life/land-management-restoration/overview.

[5] https://unu.edu/ehs/series/land-restoration-5-key-elements-reviving-our-earth#:~:text=Land%20restoration%20refers%20to%20the,and%20mitigate%20climate%20change%20impacts.

[6] https://unu.edu/ehs/series/land-restoration-5-key-elements-reviving-our-earth#:~:text=Nearly%2080%20per%20cent%20of,efforts%20for%20climate%20change%20mitigation.

[7] https://www.unep.org/news-and-stories/statements/world-environment-day-2024-accelerating-land-restoration-drought#:~:text=Every%20dollar%20invested%20in%20restoration,storage%20and%20slows%20climate%20change.

The Background and Relevance of This Book

When Saudi Arabia was the President of the G 20 countries in 2018, it allocated a substantial amount of resources for efforts towards land restoration. This has led to the creation of the G20 Global Land Initiative (G20 GLI) which is part of the United Nations Convention to Combat Desertification (UNCCD). The purpose of this initiative is not to replicate what other initiatives for this purpose are attempting, but to take an umbrella view and identify gaps and facilitate actions to bridge these gaps. Muralee Thummarukudy, an author of the book, is the Director the G20 GLI, and Santhakumar, another author, has served as a senior economist (a consultant) of the initiative.

Through an initial literature review, this team identified certain challenges of the ongoing land-restoration initiatives (and some of these are mentioned in the previous paragraphs). There was also a realisation that the effectiveness of global and national efforts towards land restoration in each country is connected to the status and efforts for human development and there is not enough recognition or understanding of possible connections between these two. There were not many efforts to link the aggregate socioeconomic factors of a country and the success/failure of efforts towards land restoration. Even when socioeconomic factors are taken into account, these were primarily based on a partial equilibrium approach or multivariate analysis (econometrics) wherein the influence of one variable (say, the prevalence of private property rights) was assessed by controlling other variables. This was found to be problematic for two reasons: First, the presence of such an enabling factor need not be effective in all contexts, and this could be due to the presence/absence of other factors. Or it could be due to the interaction between different factors; secondly, there was not enough understanding on what enable land restoration at the aggregate level in a country.

Through consultations with a set of other academics and practitioners, an approach to understand the interrelations between different factors at the level of each country, which enable or disable land restoration, was developed.[8] The G20 GLI collaborates with Ritika Jain (the second author) and her team at the Centre for Development Studies, Trivandrum (Kerala, India) to analyse socioeconomic factors which enable/disable land restoration in ten countries which have made the maximum commitment towards this goal globally.

In addition to these country-level studies, case studies were attempted to understand the actions of specific actors - private companies, non-governmental organisations and communities, towards land restoration in different parts of (mainly less-developed) world. Insights from all these studies form the basis of this book. It is not written mainly for academic readers or researchers. The purpose is to enhance the awareness and capability of millions of actors - and these may include government officials, private sector managers and social activists - to take rational, grounded

[8] This is detailed in https://practiceconnect.azimpremjiuniversity.edu.in/envisioning-a-national-system-of-land-restoration/; and https://practiceconnect.azimpremjiuniversity.edu.in/envisioning-a-national-system-of-land-restoration-part-2/.

and pragmatic steps towards restoration. This is also based on a realisation that the absence of such an approach may lead to a wastage of substantial amount of resources which are devoted for this purpose. Hence, the book provides a basic understanding of the fundamentals of land restoration, from a socioeconomic analysis of the factors leading to land degradation, to managing trade-offs and enabling factors and key stakeholders' perspective on land restoration.

The Organisation of the Remaining Parts of the Book

The book is divided into four parts. The first part includes this chapter (introduction) and sets the global context of land restoration. The next Chap. (2) discusses in detail the global and national efforts towards land restoration and the persisting challenges in this regard. Some of these global efforts include Bonn Challenge (2011); Land Degradation Neutrality of the UNCCD (2015); National Biodiversity Strategy and Action Plan (2017); Decade for Ecosystem Restoration (2021–2030), Great Green Wall; and also G20 Global Land Initiative. In addition, the efforts towards conserving biodiversity and addressing climate change have the potential to restore land and terrestrial ecosystems. Most countries have also come out with national commitments and plans towards restoring land. There are also assessments on the progress and effectiveness of some of these efforts. This chapter takes note of these efforts and challenges mostly through a review of literature (including reports) in order to contextualise the arguments of the book. There is a certain progress in term of land restoration or there are a few optimistic trends in this regard: There is a growing awareness on the need for land restoration. All actors including those in private sector are demonstrating an interest in land restoration; these are leading to interesting and notable practices. However there is a slower progress and much more needs to be attempted by considering the land degradation, that is going on currently, and the fact that the ongoing efforts towards restoration could not the reverse the trend of degradation. There are not enough actions on the ground in poorer and less-developed countries. There is a need for global coordinated action for land restoration.

This context-setting is followed by an analysis of socioeconomic determinants of land degradation in different countries of the world in Chap. 3. It is based on the data on land degradation compiled by the UNCCD for 115 countries.[9] The analysis indicate that the dependence on agriculture, per-capita income and inequality are associated with land degradation. Specifically, countries with a higher share of land being used for agricultural purposes are also countries that experience higher land degradation. Richer countries that have low degree of inequality witness low degree of land degradation. Low levels of human development accelerate the harmful effects

[9] We have collected information on a few more countries from the national departments, alternate databases and past studies. Thus, the final data spans across 124 countries.

of inequality and agricultural dependency on land degradation. This chapter indicates the need for analysing socioeconomic enablers or disablers at the level of countries.

Chapters 4 and 5 are devoted to analysis of socioeconomic factors in a set of five countries. Chapter 4 deals with India and South Africa, and Chap. 5 discusses Brazil, Australia and China. There are interesting differences between these major countries. The higher per-capita land-size and the control of a major share of land by the government are important enablers in Brazil whereas the cost/difficulty of enforcing public ownership over vast stretches of land is a challenge there. The small size of the average land-holing and higher dependence on land (and a relatively lower human development) are the challenges in India whereas (historical) state ownership of forests is an enabling factor. Though forests were degraded drastically in China, industrial/economic development during the last five decades has enhanced the incentives and capabilities to adopt land restoration policies. Though Brazil and South Africa have relatively higher per-capita incomes, these do not translate adequately into investments and action for better quality environment (including those for restoring land). The fact that the benefits of economic development were mainly derived by a section of society, and others were not equal participants of the process of development, were the main reasons. The long standing economic development and policies which ensured that the majority of its people benefit from such development, combined with reasonably functioning institutions and governance could keep land degradation relatively lower in Australia.

Part II carries a detailed analysis of the trade-offs and incentives and disincentives of land restoration. It starts with Chap. 6, which has an outline of a framework for analysing socioeconomic factors at the level of a country. As noted in Chap. 3, different socioeconomic factors may impact land restoration. However the impact of one may depend on others, policies/institutions together and also the development context of the country. This warrants a systems approach to understand the enablers of land restoration within each country. In order to do so, it draws on the framework of 'national innovation systems' that is used to understand the differences in the degree of innovation in different countries. Such a system of innovation consists of 'the elements and relationships which interact in the production, diffusion, and use of new, and economically useful, knowledge … and are either located within or rooted inside the borders of a nation-state' (Lundvall, 1992). Following this approach, this chapter looks at the parts and relationships which interact in the efforts towards restoring land in a country in an aggregate manner.

Chapter 7 highlights the need to combine ecological science with socioeconomic priorities in designing the strategies and projects for land restoration. There is a consensus among scientists about what is an ideal land-restoration practice. However this cannot be accomplished in all stretches of land in a country. Land will have multiple uses but there is a need to have non-degrading use of land in all locations. There may be a need to focus on different levels of restoration. For example, though biodiversity conservation is important, the whole land of a country cannot be used for conserving biodiversity. The marginal gain from conserving same set of species in additional area may decline when it is conserved already in sizeable stretches of land. Marginal cost of biodiversity conservation could increase (as more land is conserved for this purpose) due to the need for land for other uses. Hence biodiversity needs to be conserved in specific but not all locations. There can be an improvement in carbon sequestration even if biodiversity is not conserved (say in plantations). Enhancing carbon sequestration (without biodiversity conservation) is useful in specific locations. Plantations of one or a set of species can restore land if the wood/timber is harvested sustainably. There can be restoration of land even if it is used for cultivation. Hence one standard in all locations can be problematic if people cannot adopt it for practical reasons.

Chapter 8 is about the possibilities and challenges of different financing mechanisms for land restoration. The financial resources for restoration can be private, either from individuals or companies. Both can be for-profit, altruist or with combined goals. Then there can be different kinds of altruistic organisations which are investing in projects of land restoration. There are also public sources of finance which may include grants and loans, subsidies, and those which use market instruments like the payment for ecosystem services. There are also different kinds of finance available internationally. Those available for protecting biodiversity and addressing climate change are important under this category. However the chapter notes different kinds of challenges in ensuring that money from these different sources actually reach and is used at the ground level to restore land. These include (a) Governments in poorer countries may not have enough resources (and have other needs); (b) Individuals may not be interested in land restoration as it has an element of positive externality and hence those who invest cannot gain the full benefits; (c) Transaction costs are higher for a large number of small projects and such projects are common for land restoration; (d) Many actors do not have enough capacity to access carbon and biodiversity-finance; (e) though private finance is available, its size is still small, and it may go to locations where institutions are reasonably functional, which may not be the case in many less-developed economies; and (f) Effectiveness of international altruistic investments may be lower in certain contexts (mainly poorer countries).

Though the benefits of land restoration are known, there is not enough reckoning of the costs of restoration. That is the focus of Chap. 9. There are direct and indirect costs of restoration. The latter include opportunity costs, which include the foregone benefits when land is not used or used only in specific ways, to ensure restoration. These foregone benefits may vary depending on the social context. For example, it could be very high to those people who do not have other sources of income to buy those products, which they have been getting traditionally from land (say food-grains), from markets. The cost of not using land could be lesser to those who have non-agricultural sources of income. An understanding of the cost will help us to identify the net-losers of restoration projects and design appropriate compensation programs. These compensations are necessary to ensure the effectiveness of restoration projects.

Chapter 10 deals with the distributional issues of land restoration. The costs and benefits of restoration could be unequally distributed. The case of Adivasis (indigenous people) in terms of forest conservation is discussed in the chapter which has dealt with India. The need for local people to bear a higher burden to prevent the burning of peatlands is noted in the case of Indonesia. These distribution patterns can lead to political economy challenges for land restoration projects. These may require appropriate compensation strategies on the part of governments and international organisations.

Part III discusses the role of different stakeholders in land restoration. It starts with individuals in Chap. 11 and here the relevance of behavioural change is discussed. This is important in the contexts of developments in behavioural and neurosciences which demonstrate that actions of human beings can be driven by rational and not so rational reasons. The implications of both these for land restoration need to be understood. At one level we have seen that the positive externality associated with land restoration may discourage rational individuals from taking actions in this regard adequately, and that may require public/social support/actions. However the possibility of not-so-rational actions may work in two ways. People may not take adequate actions towards land restoration even though it may be rational for them to do so. Replanting of trees after cutting it down either due to inertia, and/or an undervaluation of future benefits could be examples of this kind. On the other hand, public campaigns which encourage land restoration should not be based solely on rational grounds. There can be emotional and moral grounds on which people may take actions towards land restoration. In that sense, the use of faith groups or religious gatherings to facilitate land restoration is justified on the basis of behavioural insights. These different routes are also important while noting the growing trend of impact investing. These investors also may be driven by both rational and not-so-rational reasons. Understanding their different motivations may lead to a more realistic understanding of the possible impact of their investment strategies on land restoration.

The role of communities is the focus of Chap. 12. Communities that have managed common lands sustainably are undergoing social transformation in different parts of world. Youngsters are getting educated and/or migrating to cities within or outside their countries. Their remittances have become a major source of income for those who remain in villages. There is an increasing dependence on goods sourced from distant markets for personal consumption. The mutual dependence among community members has declined in many contexts. These can possibly lead to unsustainable management of common property resources by communities. As part of the work of the G20 GLI, we have seen such issues in the management of grasslands in Central Asia and also by the indigenous communities in the Indonesian part of Borneo island. The possible decline of the ability of communities to manage common lands sustainably may warrant newer institutional arrangements. In many cases, governments have stepped in but the higher cost of public enforcement (and its associated inefficiencies) may create newer set of problems.

Chapter 13 focusses on the role of Non-Governmental Organisations in land restoration. There are a number of non-governmental organisations (NGOs) involved in land and ecosystem restoration. However, there is not enough systematic analysis of the potential and limitations of NGOs in this regard. Such an analysis should assess the ideological positions and operational plans and projects. NGOs have an important role in building innovative pilot models of restoration and encourage the government to act where it is not adequately responsive to the issues of land degradation. However, certain ideological positions of NGOs are not that enabling to achieve scaled up restoration. Moreover, there may be challenges in learning appropriate lessons from the innovative work of NGOs and using these for macro-plans and actions for land-restoration.

The emerging role of private companies is discussed in Chap. 14. Conventionally, the activities of private companies might have caused land degradation. This trend continues in certain contexts and that may require government regulations and their enforcement. However currently a set of private companies views restoration as a viable business for financial, social and environmental considerations. These firms work in different parts of not only the developed but also less-developed world. This chapter analyses the possibilities and limitations of these firms. An important observation is the challenges in scaling up their operations. The chapter also speculates on the possible ways to overcome these challenges.

Given the potential and limitations/challenges of other stakeholders like communities, non-governmental organisations and private companies, the role of governments in enabling land restoration is a lot more complex. Identifying and outlining these challenges is the goal of Chap. 15. First, it should facilitate the actions by other stakeholders which are socially beneficial. Secondly, it should take actions which others may not be able to take towards land restoration. Thirdly, it should regulate those activities of people and firms which may cause land degradation. Though the literature on land degradation touches upon a number of actions and omissions of governments, a comprehensive and rational view of what governments may attempt is not that explicitly available. The development of such a view is the goal of this chapter.

Given the understanding of the role of different stakeholders, Chap. 16 describes the case of peatland restoration in Indonesia. This demonstrates the challenges when different stakeholders and their (dis) incentives interact. The incentives of local people (even in considering their possible environmental losses) may not enable peatland restoration. However, global environmental losses due to the degradation of peatlands are substantial. Hence, there is a need for global action considering that it would be difficult for a country like Indonesia to bear the full costs of mitigative actions. This is especially so when the preservation of peatland yields only a smaller part of the benefits to, and the cost (including the opportunity cost) of conservation is substantially higher for, the people of Indonesia.

Part IV outlines the way forward. Education has to be an important instrument for land restoration. It is discussed in detail in Chap. 15. Land restoration also requires enhancing the awareness of all people but also the capacity of professionals and policy-makers. What kind of education and training programs may be launched for this purpose? An answer to this question has to be based on the understanding of the challenges in using education (especially higher education) for one or other social purpose (as discussed in Shephard and Santhakumar, 2023). This chapter outlines the kind of education programs (for youngsters and for adults and as part of lifelong learning) which may be effective in enhancing the quantity and quality of human resources for land restoration.

Based on the discussions in other chapters, final Chap. 18 summarises the key observations of the book. It also speculates on the kind of policies and approaches which are required at the national and international levels to facilitate restoration. The focus is on connecting initiatives which aim at reducing land degradation with other development objectives (like those which are aimed at Sustainable Development Goals) and also those which aim at addressing climate change and biodiversity. It also speculates on the possible role of private business and different kinds of investment trends on land restoration.

Chapter 2
Land Restoration: Global Efforts and Persisting Challenges

International and national actors are aware of the need to avoid land degradation. Hence, there are a number of global and regional initiatives for this purpose. These include those aimed at biodiversity conservation and addressing climate change (Mappin et al., 2021). There are also commitments to restore land (and forests) by almost all nations. Earlier milestones in this regard are the Global Partnership on Forest and Landscape Restoration in 2003, the Bonn Challenge in 2011, which set ambitious targets for land restoration, and the New York Declaration on Forests in 2014, which strengthened the resolve to conserve forests worldwide. Then the UN system has declared 2021–2030 as the UN Decade on Ecosystem Restoration. The G20 Global Land Initiative is the recent international initiative, started by Saudi Arabia (when it was the President of the G20 countries in 2020) but it has a global reach. In addition there are regional initiatives like Great Green Wall, Middle-East Green Wall, etc. We discuss these initiatives in a little more detail and the persisting challenges of land restoration in this chapter.

Land Degradation Neutrality by 2030

The UNCCD's land degradation neutrality (LDN) aims at halting, and then reversing, land degradation. It notes that 131 of the world's 196 countries have pledged (or are aiming) to arrest land degradation by 2030.[1] The actions required for this purpose are: '(a) avoiding new degradation of land by maintaining existing healthy land; (b) reducing existing degradation by adopting sustainable land management practices that can slow degradation while increasing biodiversity, soil health, and food production; and (c) ramping up efforts to restore and return degraded lands to a natural or

[1] https://www.unccd.int/land-and-life/land-degradation-neutrality/overview.

V. Santhakumar et al., *Land Restoration: A Rational, Grounded and Pragmatic Approach*, https://doi.org/10.1007/978-3-032-26652-1_2

more productive state'. However, the actual impact of LDN is yet to show up as a reduction in land degradation in different parts of the world.

Bonn Challenge Initiative

In 2011, the German government and the International Union for Conservation of Nature (IUCN) jointly initiated the Bonn Challenge, which set a target of restoring 150 million hectares of land by 2020. This goal was extended to 350 million hectares by 2030 (Verdone et al., 2017). There were also periodical assessments of the achievements of the Bonn Challenge Initiative carried out by IUCN. One report in 2019 notes that 'the United States has not only achieved but exceeded its pledge to bring 15 million hectares into restoration by 2020, reaching 17 million hectares as of June 2019. Overall, the five countries assessed in depth—Brazil, El Salvador, Rwanda, Mexico and the U.S.—had made commitments totalling 30.7 million hectares with 89% (27.385 million hectares) under restoration'.[2] Another set of 14 countries, namely India, Uganda, Malawi, Kenya, Burundi, Cote d'Ivoire, the Democratic Republic of the Congo, Mozambique, Cameroon, Ghana, Costa Rica, Guatemala, Sri Lanka, and Colombia have progressed partly. Altogether, these 19 countries have collectively pledged a total of 97 million hectares, representing 57% of current commitments. The current picture (while writing this chapter) is that around 210 M Ha are pledged.[3] It may be noted that these are only commitments and the actual progress in terms of restoration could be much less.

UN Decade of Ecosystem Restoration

This is a joint initiative of the Food and Agriculture Organisation (FAO) and the United Nations Environment Program (UNEP). Its objective is to prevent, halt and reverse the loss of nature. It has several sub-plans like that for the Forest Restoration, Restoring and Regenerating Polynesian Forests, Sahel Integrated Resilience Program, etc. However the areas that will be covered under such specific programs may be between a few hundreds to hundred thousands of hectares. The total expected impact as on early 2025 is on less than 2,000,000 H.[4]

There are a few regional initiatives such as.

AFR100 (African Forest Landscape Restoration Initiative): The goal is to restore 100 million hectares of African land by 2030. This initiative encouraged a number

[2] https://iucn.org/news/forests/201907/report-captures-achievement-us-bonn-challenge-pledge-restoration-progress-19-countries.

[3] From the website https://www.bonnchallenge.org/.

[4] https://www.decadeonrestoration.org/.

of African countries to commit to restore a part of their land. A report in 2023[5] notes: It is implemented at the national level by 34 African governments in collaboration with 39 technical and 14 financial partners. Currently, these 34 governments have committed to restore 129.5 million hectares of degraded forests and lands, implying that the original commitment has been exceeded by 29.5 million hectares. However, the initiative faces a number of challenges to attract enough investments. These include: 'A. Lack of a communication strategy in AFR 100 implementation to inform governments, the private sector and financial institutions on FLR successes and impacts. B. Competing demands at government level due to availability of limited national budget and political pressure from other sectors. C. Bureaucratic processes within government ministries. D. Governance issues: land and tree tenure.'

Initiative 20x20 in Latin America and the Caribbean: It started in 2014 with a goal of restoring 20 million hectares of land by 2020. Currently, 18 countries and 3 regions have committed to restore 52 million hectares by 2030. It is reported that 8.2 million hectares of degraded land are undergoing restoration efforts. A notable point is that it could attract 'a coalition of impact investors and funds deploying US$3.09 billion in private investment'.[6]

ECCA 30 (Europe, Caucasus, and Central Asia 30): It aims at restoring 30 million hectares of forest by 2030. It was launched in 2019. There was a ministerial meeting in 2021 to enhance the political commitment to the initiative. It is too early to have a review of the achievements of this initiative.

The Great Green Wall Initiative: The plan is to restore 100 million hectares of degraded land. It has secured $19 billion in pledges. The UNCCD launched a campaign 'to inspire a global popular movement to deliver this urgent African-led dream by 2030'.[7] Campaigns[8] provide examples of success. These 'include more than 50,000 acres of trees planted in Senegal'. Most of these are the acacia species *Senegalia senegal*, which has economic value as it produces gum arabic. (Gum arabic is primarily used as a food additive). One report notes: Thus far, progress has been made in restoring the fertility of Sahelian lands, including the planting of more than 18 million trees in Senegal, eight million in Nigeria, 129 million in Eritrea, and 146 million in Niger.[9] A small portion of these trees are also fruit-bearing, which, when mature, will help combat the high levels of malnutrition in the rural interior.' However, the real achievement so far is only a very small part of the goal.

[5] https://afr100.org/sites/default/files/2023-11/SUMMARY%20OF%20FOREST%20LANDSCAPE%20RESTORATION%20SUCCESSES%2C%20CHALLENGES%20AND%20SUGGESTIONS%20FOR%20HIGHER%20MOBILIZATION%20OF%20INVESTMENTS%20TO%20ACCELERATE%20AFR100%20IMPLEMENTATION%20IN%20AFRICA.pdf.

[6] https://initiative20x20.org/restoring-latin-americas-landscapes.

[7] https://www.unccd.int/our-work/ggwi.

[8] https://education.nationalgeographic.org/resource/great-green-wall/.

[9] https://www.iisd.org/articles/deep-dive/land-degradation-land-restoration.

The Middle-East Green Initiative: This Initiative was launched in 2022 by Saudi Arabia with an objective to plant 50 billion trees to restore 200 million hectares of degraded land. To accelerate the implementation of these goals, Saudi Arabia allocated 2.5 billion US dollars to support projects and governance. It is too early to get signals of success, but the allocation of a substantial amount of resources may make a difference.

G20 Global Land Initiative

The G20, or Group of Twenty, is an intergovernmental forum of the world's major developed and developing economies. It comprises 19 countries (Argentina, Australia, Brazil, Canada, China, France, Germany, India, Indonesia, Italy, Japan, Republic of Korea, Mexico, the Russian Federation, Kingdom of Saudi Arabia, South Africa, Turkey, the United Kingdom, the United States of America), and the European Union. Collectively, the G20 accounts for 85% of global GDP, two-thirds of the world population and make up about half of the earth's total land area. The ambition of the G20 Global Initiative on Reducing Land Degradation and Enhancing Conservation of Terrestrial Habitats is to achieve a 50% reduction in degraded land by 2040. The Initiative stresses that ecosystems including, but not limited to, forests, grasslands, drylands, rangelands, croplands, peatlands, mangroves, soil, tundra, karst and wetlands, are essential for the existence, survival and sustainability of biodiversity as well as food security and human well-being. Therefore, the Initiative will cover the entire range of ecosystems, which also include riverine systems and watersheds. The initiative will work not only with governments but also with all other stakeholders: the private sector, academia, the media, communities, non-governmental organisations, faith-based institutions, financial institutions, etc. It is important to note that countries make restoration commitments voluntarily in this initiative. The Initiative will establish a number of new platforms to bring together the stakeholders who are interested in the domain area of land restoration and develop strategies and action plans to increase their participation. It is too early to make an assessment of the impact of the initiative. This book is an outcome of the efforts of G20 Global Land Initiative. The impact of these activities on its goal (to achieve a 50% reduction in degraded land by 2040) is unpredictable. However, there could be positive benefits in the long run. A deeper understanding of the socioeconomic enablers of land restoration, and informed actions to create these enablers (the purpose of this book) could be an outcome with such a positive benefit.

Land Restoration as a Strategy Towards Conservation of Biodiversity

The potential and limitations of land restoration towards the conservation of biodiversity are noted in the literature.[10] The dominant approach to conserving biodiversity so far is to preserve pristine areas. However, there could be scarcity of such areas in the future, and hence there is a need for a refinement of restoration techniques which facilitate the sustenance of a higher level of biodiversity. However, currently efforts towards land restoration in most parts do not achieve the goal of restoring native biodiversity (Edwards and Creole, 2024). The conservation of biodiversity may not be the objective of all projects of land restoration (which we discuss in the following chapter). However, there are possibilities of restoration of habitat on derelict sites, but also techniques for enlarging and redesigning existing reserves. Restoration may even make it possible to move reserves entirely in response to long-term events, such as changes in climate (Simonson et al., 2021). It also provides opportunities to increase diversity through activities as diverse as management of utility corridors, transportation rights-of-way, and parks (Francois et al., 2024). Though all efforts towards land restoration may not conserve biodiversity, the reverse may happen. The efforts towards the conservation of biodiversity can contribute to land restoration. There are commitments on the part of governments and the private sector for the preservation of biodiversity. An example of the latter is the foundation for biodiversity.[11] 'A total of 194 financial institutions representing 29 countries and over 23 trillion euros in assets signed the Finance for Biodiversity Pledge in 2023. The Pledge was initiated by a group of 26 financial institutions calling on global leaders and committing to protect and restore biodiversity through their financial activities and investments and launched during the Nature for Life Hub on 25 September 2020 and the Biodiversity Summit of the United Nations General Assembly on 30 September 2020.'

Land Restoration as a Strategy Towards Addressing Climate Change

Processes of land restoration and rehabilitation, and the restored land, have tremendous potential to absorb CO2 and limit climate change (Devi et al., 20,240.[12] It is argued that the restoration of 15% of converted land in priority areas could sequester 299 billion tonnes of CO2 (Strassburg et al., 2020). This is about 30% of the total atmospheric CO2 increase since the Industrial Revolution. However, different restoration ways (such as protecting, managing and restoring land) may

[10] For example,Jordan et al., (1988) and Rapiya et al., (2024).

[11] https://www.financeforbiodiversity.org/.

[12] IRP (2019).

differ in terms of the potential contribution towards mitigating climate change and there is a need to take these into account in prioritising strategies (Cook-Patton et al., 2021). Though land and ecosystem restoration has huge potential, only 1%of finances globally are directed towards this purpose.

Indicators of Progress in Land Restoration

The most important sign of progress is that there is a wider recognition of the problem of land degradation and the need to reduce it. This is evident from national commitments which together have promised to restore one billion hectares, an area larger than China.[13] This enhanced awareness can be seen in both developed and developing countries. The initiatives which we have mentioned are partly causes of national actions, and in certain cases, the outcome of an increasing awareness. One can see a certain transformation in the farming practices in many countries (mostly developed countries) which combine the goals of farming and restoring land.[14] The last four decades have also seen the arrival of many useful technical, social and policy practices, mainly in the developed world. The private sector is showing greater interest in land restoration as a business opportunity, and that is another positive trend.

Regarding the status of land degradation in different countries, one report[15] has given the following picture: 'One hundred fifteen countries have publicly available data on the UN SDG Database for both the years 2015 and 2019. Of those 115 countries, 41 show progress in reducing land degradation between the two data points. Two countries show no change: Poland with 0.2% land degradation and Lesotho with 8.83% land degradation. Seventy-two countries show worsening land degradation between the two years'.

Persisting Challenges

The non-achievement of deadlines for such international initiatives for ecosystem restoration is not uncommon. The Target 16 of the Convention for Biological Diversity had the goal to rehabilitate at least 15% of degraded ecosystems by 2020. However, a mid-term review noted that many countries were unlikely to meet these targets by the deadline (Farhadinia et al., 2022). This is mainly due to a number of challenges, and some of these are mentioned in the following paragraphs.

---

[13] https://www.unep.org/news-and-stories/statements/world-environment-day-2024-accelerating-land-restoration-drought.

[14] For a description of the situation in Australia, refer Campbell et al., (2017).

[15] https://www.landesa.org/wp-content/uploads/Brief-Progress-towards-the-SDG-land-degradation-and-restoration-commitments-2023.pdf.

Not Enough Financial Resources

It is noted that activities which may cause harm to nature receive around 7 trillion USD, whereas that allocated for its restoration is relatively minuscule (UNEP, 2023). Though countries make commitments towards land restoration, the resources which are allocated for this purpose are grossly inadequate. There is a notable gap between the goals and reality in restoration projects even in developed countries (Collard et al., 2019). It is noted that to avoid additional land degradation and to rehabilitate 12 million hectares of degraded land each year, traditional financial sources, official development assistance, and national government resources alone will not suffice (Chasek et al., 2019). This indicates the need for private finance towards land restoration. However, the global financial system is also not designed appropriately to meet the long-term goals like land and ecosystem restoration (Garvey et al., 2025). Even when private finance moves to land restoration it may become available for certain solutions and certain regions. One study notes that 'investment finance biases towards restoration projects in low-risk areas and corporate finance towards areas with business presence. Both private finance types tend to omit projects focussing on natural regeneration' (Lofqvist et al., 2023a).

Though the imperatives of addressing climate change helped create 'climate finance', and there is an expectation that it will make a substantial amount of resources available for activities such as land restoration, the actual amount of resources which reach the ground for restoration from this pool, especially in less-developed countries, is limited. This could be partly due to the institutional and technical difficulties in accessing these resources. The actors, including government officials in developing countries, may need a higher level of capacity for this purpose.[16]

Limited funding for land restoration can be due to competing priorities for public financial resources. For less-developed countries, there are competing demands for public resources. Moreover the cost (including opportunity cost) of land restoration can be higher, especially where people have a higher dependence on land. The involvement of people and communities is important to have effective land restoration. Even when there are international and national actions and resources, the effectiveness of land restoration depends on private individuals and households, especially those involved in agriculture. There are two constraints in this regard. First, their private incentives may not be in tune with restoration. Secondly, they may not be aware of the need for, or have behavioural patterns which are in tune with, land restoration. Changing both these requires intervention by governmental and non-governmental organisations. But the actions of government in democracies are determined by individuals, and hence this can create a vicious cycle.

[16] We have noted this in the case of peatland restoration in Indonesia.

Small-Scale Operations of Land Restoration as a Business

There is a budding interest among private companies to see restoration as a business. It is noted that restoration can be a potentially sizeable market. However, as part of the G20 Global Land initiative, we have seen the operations of some of these organisations. Though there are laudable efforts, these could not achieve a scale so as to have a significant impact on reducing land degradation. Most of these efforts face difficulties in accessing higher amounts of capital and scaling up their businesses.[17] Even in countries where there is a notable traction of private industry in land restoration, there are institutional deficiencies which affect their activities.

Inadequate Capacity in Less-Developed Nations

Though there are national commitments, these may not translate into actions if there is no building of capacity and institutions at the national level. It is noted that the main agency at the national level lacks the personnel and knowledge to coordinate integrated land use planning and LDN implementation.[18] Other challenges include the lack of scientific knowledge and data to develop integrated land use planning in some countries, uncertainties about the influences of global drivers like climate change, poor coordination between national governments and local authorities, and lack of knowledge and systems for enforcing, monitoring, and evaluating LDN measures (Chasek et al., 2019).

Contradictory Goals of Governments

There could be contradictions between development and restoration goals. One way countries encourage poverty reduction is to provide subsidies in agriculture. This may encourage farmers to use farm inputs intensively to produce more outputs. This can cause land degradation. It is possible to provide subsidies for land-restoring agriculture. But this may require transfers which are higher than the profit from degrading agriculture (Shi and Chen, 2004). Many countries may see this as inadequate to address poverty eradication and may not have the higher levels of resources which are needed to subsidise restoration-enabling agriculture.

There are poorer countries which do not have enough resources even for poverty reduction. Hence poorer people may not get much resources for their basic needs. The only resource that they may have the access to some land (even though it may not be privately owned). Hence, people are compelled to use the land to meet their

[17] One such case is described here https://practiceconnect.azimpremjiuniversity.edu.in/private-businesses-in-forest-conservation-the-case-of-forestwise-in-borneo-island/.

[18] https://www.iisd.org/articles/deep-dive/land-degradation-land-restoration.

needs for food (and it may not be enough to provide adequate nutrition). This can lead to practices which can lead to land degradation.

Private companies use land for multiple reasons. Land can be an important input into production, as in the case of mining companies. Though certain degradation of land is inevitable as part of the process, there can be practices which can minimise this degradation. However, this may require regulations which compel the industry to carry out its activities with minimal degradation of land. However, many countries, though they have regulations, don't enforce these strictly. Hence one can see degradation continuing as part of industrial activities. Governments may not be able to enforce regulations tightly even if these exist. This could be partly related to the scarcity of public resources for the adequate enforcement of regulations. It can also be due to the weakness of institutions (rules of the game). There could be collusion between enforcement agencies and companies which degrade land. To some extent, this is also connected to the issues of human development, and how citizens demand better governance from their governments.

Urbanisation Is an Unplanned Process in Many Contexts

Urbanisation in developing countries is an unplanned process. Moreover, poverty and other forms of deprivation in rural areas encourage people to move to cities in search of a livelihood. This will aggravate the development of human settlements in peri-urban areas[19] with minimal public facilities. The regulations related to land use (which may be driven by the need to address the vulnerability of sections of society) may also complicate the process of urban growth. All these can lead to urbanisation which causes land degradation.

Lower Levels of Human Development

Despite all the focus on economic and human development in global discourses, many countries, especially those in South Asia and Sub-Saharan Africa, have not achieved desirable levels of human development. This is reflected in the levels of poverty, lack of education and ill-health. Large sections of people have not been able to get a decent income from industrial and service-sector activities, which depend less on land. Moreover, they may not have access to cheaper food through governmental transfer systems, and hence survival depends on extracting whatever can be cultivated from the land, even if it is not in a healthy condition. This can be an important reason for the persistence of land degradation in these parts of the world.[20]

---

[19] This issue is discussed in a number of articles like Salvati and Zitti, 2007 and Egidi et al., (2020).

[20] The relationship between human development and land degradation need not be uni-directional. The degradation of land and ecosystem can affect human development too. For example, this

Private actions need not be in tune with social and environmental goals. There can be contradictions between private incentives and public policies.[21] The possible connection between agricultural subsidies and land degradation is well discussed in the literature.[22] However, these subsidies aim at addressing certain social or developmental problems. Or the removal of subsidies becomes difficult due to a political economy which is shaped by the vulnerability of those people who depend on land.

Reports have also noted the lack of coordinated actions in different sectors like agriculture, forestry, fishery, etc. (UNEP, 2023). The lack of integration between environmental goals and other development goals is also noted as a major challenge (UNEP, 2023). Even when international efforts and resources are available to restore land in less-developed countries, there could be a conflict between the incentives of people within these countries and those of global actors. This was found to be working against the effectiveness of peatland restoration in Indonesia.[23] People who live in peatland areas, though they can gain from restoration, their cost of not using this land (or of using it in manners which conserve it) is costlier. Hence such incentives or disincentives at the local and national levels may impact the effectiveness of global restoration efforts. This book is a modest attempt to discuss these challenges in detail. It is expected to contribute to actions towards large-scale land restoration in different parts of the world.

Conclusion

This chapter notes that different actors of the world have considered land degradation as a major challenge, and have designed different initiatives to arrest it at the international, regional and national levels. However the implementation of these initiatives is not moving along the expected pace, and much more needs to be attempted to arrest the ongoing land degradation. Incentives of individuals (farmers) and firms are not in tune with land restoration. Many people are compelled to use land in a degrading manner. Many countries and their governments may not have enough public resources to support land restoration. Though there is a growing interest in land restoration as a business opportunity among sections of private investors, these may get limited to certain social contexts. All these indicate the need for an analysis of socioeconomic factors which enable/disable land restoration and that is attempted in the next chapter.

is seen in Yemen. https://www.undp.org/yemen/publications/land-degradation-and-human-development-yemen.

[21] Many such issues are noted in United Nations Convention to Combat Desertification, 2022. The Global Land Outlook, second edition. UNCCD, Bonn.

[22] https://www.wri.org/research/farm-restoration-subsidies.

[23] https://practiceconnect.azimpremjiuniversity.edu.in/peatland-restoration-in-indonesia-needs-a-global-collaborative-and-pragmatic-approach/.

Chapter 3
Socioeconomic Determinants of Land Degradation Across Countries

The role of socioeconomic factors in land degradation is recognised theoretically.[1] There are empirical studies that have assessed the role of different socioeconomic factors on land degradation. Econometric exercises analysed farm-level factors of land degradation, and these could see the impact of socioeconomic factors such as poverty (Abu Hammad and Tumeizi. 2012; Skelincka et al., 2022). There were also comparisons between regions (agro-ecological zones) within a country, namely Vietnam (Vu et al., 2014) and Poland (Istanbuly et al., 2022). There are descriptive analyses of the role of socioeconomic factors in specific areas such as (peri-urban areas) in southern Europe (Seifollahi-Aghmiuni et al., 2022). By looking at the discontinuities at the national borders, studies have documented the systematic impact of national policies on forest conservation (Wuepper et al., 2024a) and on soil erosion rates in croplands (Wuepper et al., 2024b).

There are efforts to integrate socioeconomic variables in the analysis of land degradation (Lofqvist, 2023). There is a need to have more assessments at different levels which integrate socioeconomic and institutional aspects with driving factors of land degradation (Sigman and Elias, 2021; Chazdon et al., 2021). Additionally, socioeconomic factors also affect the responses designed and implemented by the government and other stakeholders. Thus, pre-existing socioeconomic and institutional factors need to be integrated in the framework that relates drivers and responses of land degradation. This is more important at the national level where policies for socioeconomic development and land restoration are pursued. The need for context-specific analysis of land degradation and socioeconomic factors like poverty is recognised from literature reviews (Gerber et al., 2014).

[1] One of the first theoretical analysis of the role of economic factors on land degradation is Barbier (1997).

V. Santhakumar et al., *Land Restoration: A Rational, Grounded and Pragmatic Approach*, https://doi.org/10.1007/978-3-032-26652-1_3

This chapter presents a global picture of determinants of land degradation at the country level. Specifically, we use data on a set of countries and examine the socioeconomic factors that enable or disable land degradation. Since establishing causal inference is not a straightforward exercise with land degradation, we limit our analysis to the association between land degradation and specific socioeconomic factors.

Details of the Data

Data on land degradation is collected from the UNCCD country-level reports. It has country-level data on land degradation in 2015 and 2019 (share of area degraded and size degraded) for approximately 120 countries (names are mentioned in the Annexure I of this article). These countries have submitted official reports with national-level land degradation data for SDG indicator 15.3.1 to UNCCD.

The UNCCD method of computation of the land degradation measure (SDG indicator 15.3.1) is based on combining three sub-indicators—namely, land cover, land productivity and soil organic carbon stock for each country.[2] The baseline year for the indicator is 2015 and its value is derived from an initial quantification and assessment of time series data for the sub-indicators for each land unit during the period 2000–2015. We discuss the detailed method of construction in Annexure II of this Chapter. The other reporting year is 2019 and is valid for 2016–2019.[3] Hence, we focus on the share of degraded land in 2015 and 2019 in the total country area. We also collect information on covariates that may affect degradation from multiple sources. Since land degradation is relatively a slower and structural process, we take a four-year lagged value for each of these covariates. An alternate approach is to compute cumulative land degradation up to a given year as followed by Wuepper

[2] There are issues with this data. We are thankful to an anonymous reviewer who highlighted the following: The way the UNCCD measures land degradation has well known strengths and weaknesses, compared to alternative approaches used in academic research. This must be taken into account when interpreting the results. Of course, no measure is perfect but there are two key aspects of the approach that can really lead to a misunderstanding of trends. First of all, the measure is purely quantifying trends and there is no causal attribution whatsoever, and secondly, deterioration of a single indicator classifies a location as degrading. What is the issue? Without any causal attribution of changes in the indicators and by defining land degradation as a reduction of any single indicator, the measure is extremely broad and it is very difficult to interpret it. To give just one illustrative example: Globally, intensively farmed grasslands are losing their biodiversity because many species need extensively managed grassland (not too much fertilizer, not too often mowed, etc.). The UNCCD is located in Germany, which is a country that has agri-environmental payments to reduce grassland management intensity (in order to restore its degrading grasslands!). According to the UNCCD measure, lowering management intensity is classified as land degradation, because biomass production goes down, and this is sufficient to classify the successful restoration policies as causing degradation.

[3] More details of the construction and concept of the measure is available on https://unstats.un.org/sdgs/metadata/files/Metadata-15-03-01.pdf.

Table 3.1 Details of the data

Variable	Measure	Definition	Source
Land degradation	Degraded share	Share of the total geographical area of a country that is degraded	UNCCD country level reports, SDG 15.3.1 achievements
Development	Human development index	Arithmetic mean of life expectancy, literacy and per capita income	Human development reports
Educational attainment	Attainment levels of individuals	Mean years of schooling	UNESCO website
Inequality	Gini index	The extent to which the distribution of income or consumption among individuals or households within an economy deviates from an equal distribution	World bank open data
Income	Per capita income	Log (real per capita income in constant terms US$-2010)	World bank open data
Structural transformation	Agriculture dependency	Percentage (of total land) used for agriculture	World bank open data

Table 3.2 Summary statistics of the variables

Variable/Measure	Mean in 2015	Mean in 2019
Land degradation%	13.08	15.96
Human development index	0.69	0.72
Ln (per capita income)	8.62	8.71
Mean years of schooling	8.25	8.67
Income inequality	37.95	39.15
Agriculture dependency	36.92	37.01

et al., (2021). However, since we have land degradation data on only two periods, we try a robustness exercise by increasing the lag to ten years.

Our variables and measures are compiled in Table 3.1.

UNCCD data on land degradation spans across 109 countries in 2015 and 2019. Data for an additional 15 countries have been collected from other international and national policy documents.[4] Among these 124 countries, the average level of degradation has increased from 13.08% in 2015–15.96% in 2019, implying land degradation is worsening rapidly. Summary statistics for the 124 countries are compiled in Table 3.2.

[4] Various sources include the UNCCD SDG 15.3.1 goal indicator, FAO and published academic journal articles.

Table 3.2 indicates that human development index and per-capita income have increased marginally over time. Additionally, income inequality and agricultural dependence have risen marginally. We start with some basic graphical analysis in the next section.

Basic Plots of Land Degradation with Other Variables

We start with some scatter plots of land degradation with various covariates. We have also plotted the fitted regression line in the same graphs. Each of the figures is a horizontal panel of two plots for the two years considered for our analysis—2015 and 2019.[5]

1. Development: We find a weak negative association between land degradation and HDI as observed in Fig. 3.1.
2. Income: Fig. 3.2 shows a negative association between land degradation and per capita income of a country.
3. Income inequality: We find a positive association between land degradation and income inequality (Fig. 3.3).
4. Agriculture dependence: We find a weakly positive association between the share of land used for agricultural purposes and the degraded share of land (Fig. 3.4).
5. Mean years of schooling- Fig. 3.5 indicates a weak negative association between land degradation and mean years of schooling.

Figures 3.1–3.5 present a systematic pattern between land degradation and each of these covariates. Further, each of these covariates has a similar pattern with land degradation in both 2015 and 2019. It raises the question of whether these patterns have become weaker or stronger over time. This is a pertinent question, especially with land degradation rising over time. To test that, we calculate the correlation coefficient of each of these variables with land degradation for 2015 and 2019 (Table 3.3).[6]

As expected, along with rising land degradation, the association of these variables is also stronger in 2019. To further highlight the importance of examining these socioeconomic determinants of land degradation, we test the mean differences of HDI, agriculture dependence, income and inequality for countries that had high land degradation with their low-degradation counterparts. We begin by dividing countries into high-degradation and low-degradation groups based on whether their degradation levels were above or below the global average. The mean difference tests are compiled in Table 3.4.

The above table also reiterates the relevance of the above variables in explaining land degradation.

[5] Each plot region is divided into four quadrants according to the median values of the attributes.

[6] The p-values are given in brackets- *, ** and *** indicate statistical significance at 10, 5 and 1% respectively.

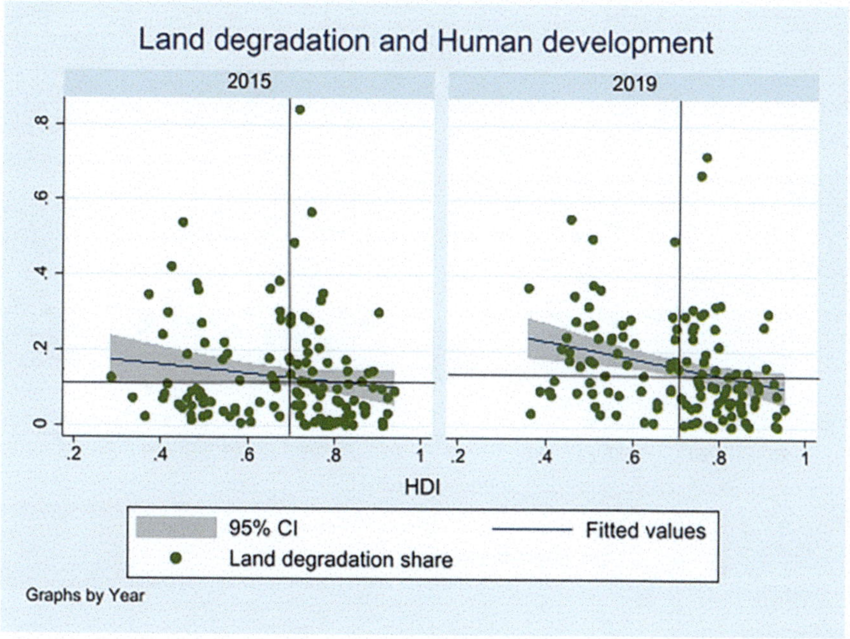

Fig. 3.1 Land degradation and human development index

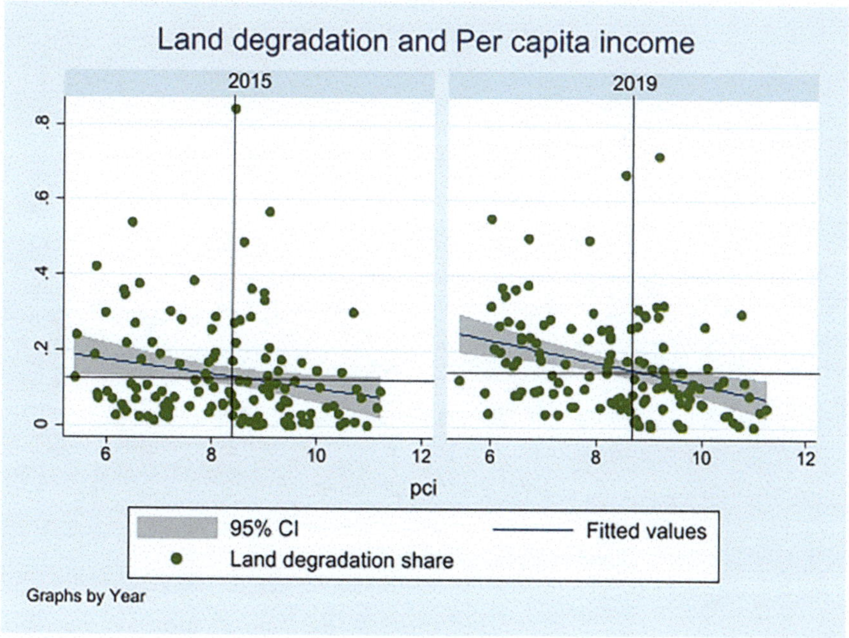

Fig. 3.2 Land degradation and per capita income

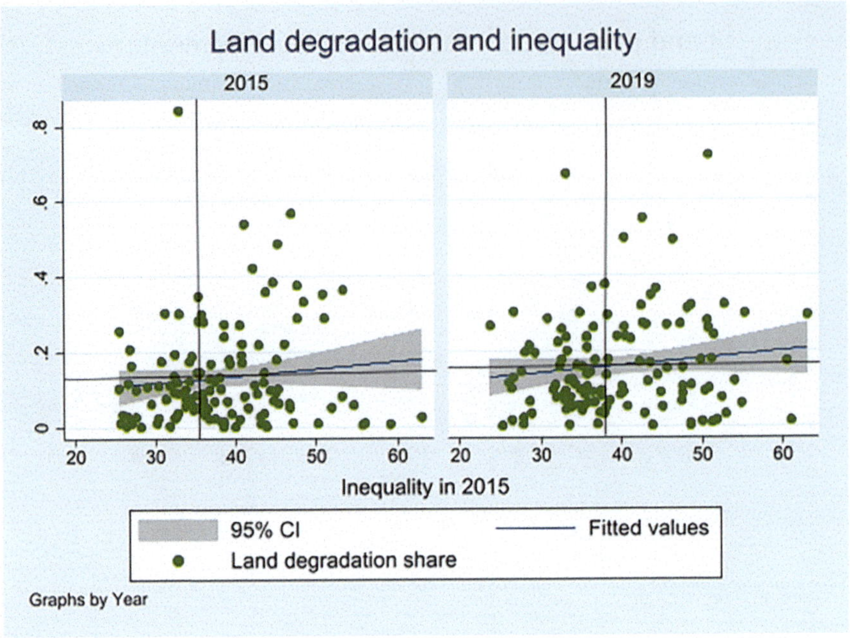

Fig. 3.3 Land degradation and income equality

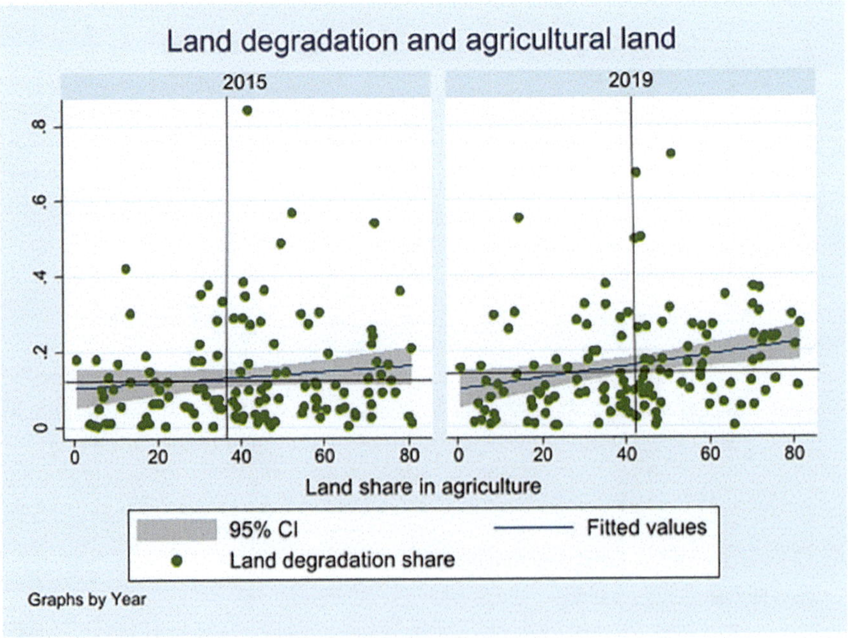

Fig. 3.4 Land degradation and share of land used for agriculture

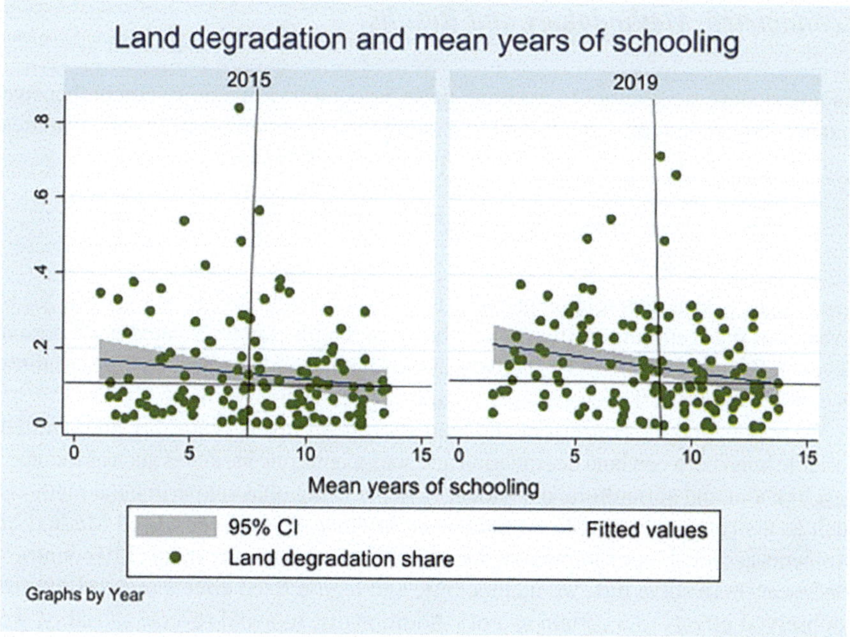

Fig. 3.5 Land degradation and educational attainment

Table 3.3 Correlation coefficient of the share of land degradation with the above variables

Specific factors	Overall sample	2015	2019
HDI	−0.194*** (0.001)	−0.143* (0.104)	−0.270*** (0.002)
Per capita income	−0.259*** (0.001)	−0.204** (0.020)	−0.330*** (0.001)
Inequality	0.117* (0.058)	0.102 (0.248)	0.119 (0.176)
Land share in agriculture	0.175*** (0.004)	0.119 (0.176)	0.236*** (0.007)
Mean years of schooling	−0.158*** (0.010)	−0.131 (0.138)	−0.204** (0.020)

Table 3.4 Mean differences of covariates across LD categories

Covariate	High LD	Low LD	Difference
Human development index	0.661	0.705	−0.043***
Per capita income (at 2015 constant prices US$)	6116.28	12301.31	−6185.02***
GINI coefficient	38.90	38.83	0.073
% share of land in agriculture	37.26	45.36	−8.097***
Mean years of schooling	8.34	7.71	0.629*

Note *, ** and *** indicate statistical significance at 10, 5 and 1%, respectively

Econometric Methodology and Results

As a next step, we attempt to examine these associations in a more comprehensive manner. We use a regression model to examine the associations of these covariates with land degradation. The regression equation is as follows:

$$LD_{ijt} = \alpha + \beta x_{ijt} + \gamma r_j + \delta T_t + \varepsilon_{ijt},$$

where LD_{ijt} is land degradation in the ith country located in jth region in year t. X_{ijt} is a vector of the confounders (discussed above) in the ith country located in jth region in year t. We include unobserved effects at the regional and year level using dummies as denoted by r_j and T_t, respectively. Finally, ε_{ijt} is the error term.

We understand the concern of endogeneity (due to reverse causality and omitted variable bias) between land degradation and socioeconomic variables such as income, development and agricultural dependence. While using quasi-experimental methods such as instrumental variable estimation or Arellano-Bond Generalised Method of Moments are ideal, our sample size does not allow us to employ them (120 countries for 2 years). To solve this, we include time and region fixed effects (controlling for unobserved effects to a certain extent). Additionally, to avoid reverse causality, we use past values of the confounding variables—four-year lagged variables in the main model.

A point worth mentioning here is that, instead of capturing development of countries using the human development index directly, we construct a dummy variable that takes a unit value if the country has an HDI score above the group average in that particular year and a value of zero otherwise. We follow this approach in Models I and II to avoid multicollinearity since we use two components of HDI- per capita income and educational attainment—as other covariates in the same model. To refine our models further, we include two more variants of our regression equation in Models III and IV. Specifically, in Model III, we include HDI (without income and education measures) along with confounding variables such as agriculture dependency and inequality. In Model IV, we drop HDI and include income along with the same confounding variables. Results are compiled in Table 3.5.

According to Model I of the above table, agriculture dependence, income and inequality are associated with land degradation. Specifically, countries with a higher share of land being used for agricultural purposes are also countries that experience higher land degradation. Further, richer countries and countries that have a low degree of inequality also share a low degree of land degradation. Since land degradation may be driven by political, institutional and policy frameworks at the country level, which may be intangible to capture and measure, we include dummy variables at the regional level to control for any unobserved effects that may impact land degradation in that region.

It is important to note that both human development index and mean years of schooling are insignificant in Model I. This raises the question of whether land degradation is driven by human development at all. Or is it just income that drives

Table 3.5 Regression results for examining land degradation share

Dep variable- LD share	Model I	Model II	Model III	Model IV
Low HDI (in model III we use HDI levels)	0.009 (0.029)	0.004 (0.123)	−0.209** (0.091)	
Share of land used for agriculture	0.0009** (0.0004)	0.001** (0.0006)	0.001** (0.0005)	0.001** (0.0005)
PCI	−0.023** (0.011)	−0.023** (0.011)		−0.023** (0.011)
Mean years of schooling	−0.001 (0.004)	−0.001 (0.004)		−0.001 (0.004)
Inequality	0.002** (0.001)	0.002 (0.001)	0.002** (0.001)	0.002** (0.001)
Low HDI # agriculture land share		0.005 (0.004)	0.008* (0.004)	0.005 (0.005)
Low HDI # inequality		0.003** (0.001)	0.002** (0.001)	0.003** (0.001)
Dummy for 2019	0.022 (0.015)	0.023 (0.015)	0.029* (0.016)	0.024 (0.015)
Regional dummy	Yes	Yes	Yes	Yes
Constant	0.238** (0.107)	0.219 (0.143)	0.161** (0.082)	0.216** (0.095)
# observation	242	242	242	242
R squared	0.146	0.144	0.120	0.139

Note: *, ** and *** indicate statistical significance at 10, 5 and 1 percent, respectively

the level of land degradation? To answer this question in a more refined way, we test if development levels condition the impact of other economic variables on land degradation in Model II.

Specifically, the interaction term of development with inequality implies that, if high inequality dampens how development affects land degradation outcomes, the coefficient of the interaction term of low human development index with inequality is positive and significant. To summarise, low levels of human development accelerate the harmful effects of inequality on land degradation. Additionally, the share of land used for agricultural purpose is positively associated with land degradation irrespective of the development levels. All other variables in Model II continue to behave in a similar way as in Model I. As expected, HDI has a negative and significant coefficient in Model III. In other words, if we include only HDI (without its components), we find low levels of development are associated with high land degradation. Additionally, low development accelerates the harmful effects of inequality and agricultural dependency on land degradation. Findings of Model IV are qualitatively similar to that of Model II.

Table 3.6 Robustness tests using ten-year-lagged independent variables

Dep variable- LD share	Model I	Model II
Low HDI	0.025 (0.032)	0.007 (0.123)
Share of land used for agriculture	0.009** (0.003)	0.001** (0.0006)
PCI	−0.019** (0.009)	−0.020** (0.009)
Mean years of schooling	−0.001 (0.004)	−0.001 (0.004)
Inequality	0.002** (0.001)	0.002 (0.001)
Low HDI # agriculture land share		0.006 (0.005)
Low HDI # inequality		0.003** (0.001)
Dummy for 2019	0.033** (0.015)	0.034** (0.015)
Regional dummy	Yes	Yes
Constant	0.148 (0.095)	0.176 (0.137)
# observation	251	251
R squared	0.155	0.159

Note: *, ** and *** indicate statistical significance at 10, 5 and 1 percent, respectively

Some Extensions

Are the above findings sensitive to the time period of our analysis? In the above analysis, we consider four-year lagged variables for all the socioeconomic determinants of land degradation. We test the robustness of our finding by considering ten-year lagged socioeconomic variables. The rationale for the specific ten-year lag is justified by how data on land degradation is collected by the UNCCD since the assessment of the baseline period ranges from 2000–2015.[7] The results are compiled in Table 3.6.

The results remain qualitatively similar. This confirms the robustness of our finding that human development conditions how inequality affects land degradation.

Is there any association between degradation changes over time and the levels of degradation?

One of the common trends that is evident in normal summary statistics and in our regression results is that there is a rise in the level of land degradation during the study period. This warrants the question if there is any relationship between the

[7] Data on the three sub-indicators for land degradation is collected across the countries in different years within this fifteen year period. We don't have the exact year in which each country had submitted its detailed report on land degradation.

initial levels of land degradation and the pace of degradation between the two years. In other words, did the countries with high land degradation in 2015 experience a faster rise in land degradation until 2019? We start with the pattern between the share of land degradation in 2015 and the increase in land degradation shares between 2015 and 2019 in Fig. 3.6. Surprisingly, we find a negative association between levels of land degradation in 2015 and the pace of degradation. This implies that countries which had lower land degradation in 2015 exhibited rapid degradation in the next four years.

To test this further, we next examine to see if countries that experienced high rates of land degradation between 2015 and 2019—also had higher shares of degraded land in 2019 (Fig. 3.7).

To summarise, we find two distressing facts about the global state of land degradation—it has risen in the short four-year period of our analysis. Further, with lower land degradation in 2015, they degraded so rapidly that they were leading in land degradation in 2019.

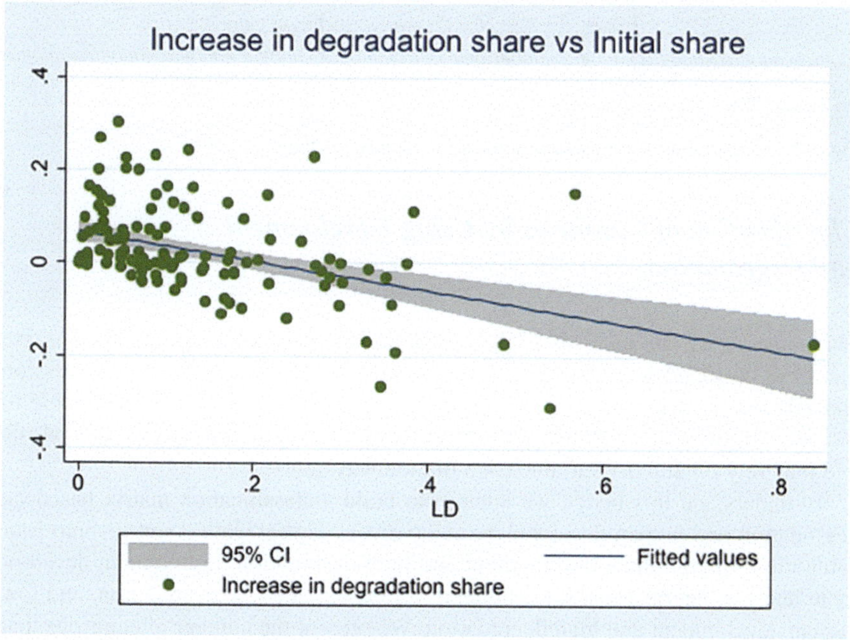

Fig. 3.6 Association of increase of land degradation with levels

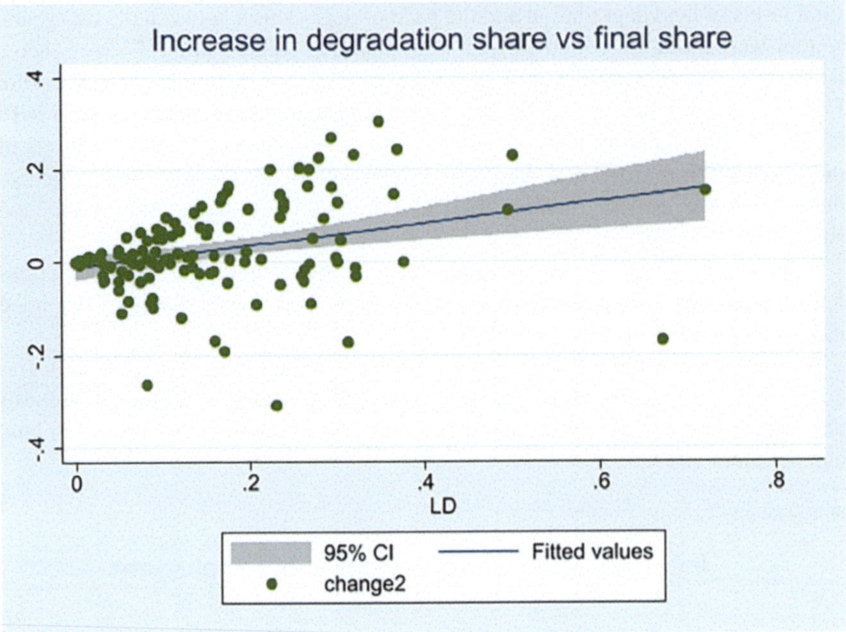

Fig. 3.7 Increase in land degradation and levels of land degradation in 2019

Classification of Countries by Using Development and Degradation

The above aggregate analysis of the global scenario provides us a comprehensive picture of how land degradation is driven by various factors. However, in a regression analysis that is based on quantifying processes and factors, we may be unable to include institutional, legislative and other policy factors systematically in our analysis and may rely on dummy variables in a limited way.

To understand this better, we attempt to build a classification matrix based on degradation and human development comparisons to segregate countries into four categories—high human development and high degradation, high human development and low degradation, low human development and low degradation, and low human development and high degradation. We present the number of countries that fall into each of the above categories in Fig. 3.8.[8]

In the above figure, the high human development and low land degradation category comprises 38 countries that have strong development and relatively sustainable land management. Further, the low HDI and high land degradation combination of 39 countries is the most vulnerable countries that are struggling in both aspects.

[8] We have included the names of the countries alongside the codes mentioned as an Annexure I to the chapter.

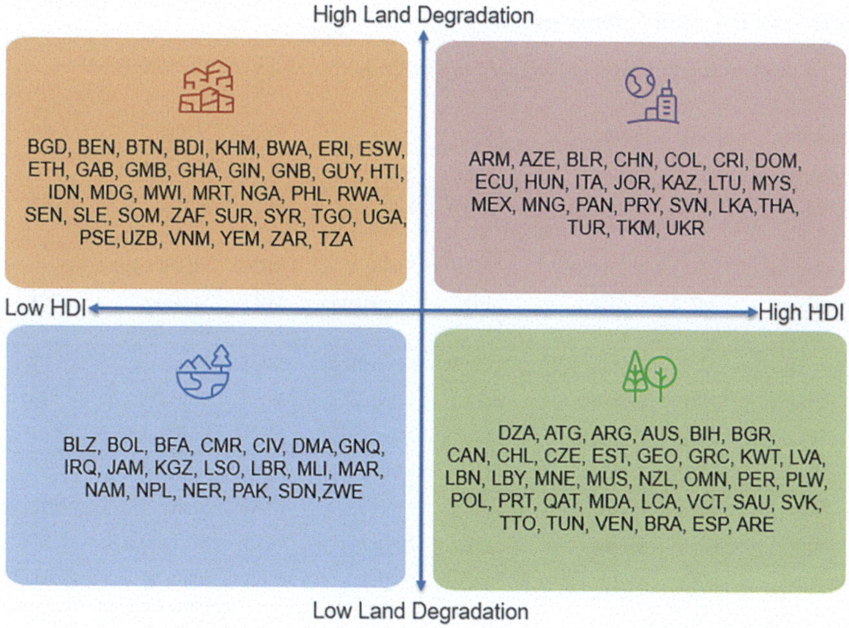

Fig. 3.8 Country categorisation according to land degradation (2019) and HDI (2015)

The above two combinations of countries exhibit an expected negative association between human development and land degradation.

However, it is the countries that fall in the other two categories that are unique and require an in-depth analysis. There are 26 countries that have high human development but with high levels of degradation, indicating stress on land, possibly due to some natural, human-induced or institutional factors. Similarly, 24 countries fall under low land degradation despite low human development. This provides an interesting backdrop, since these countries emerge as outliers as they defy the usual negative association between human development and land degradation and require careful scrutiny.

These countries require a context-specific analysis that examines the interaction of different socioeconomic and institutional factors that may be driving these differences but haven't been directly controlled in the regression analysis. In fact, differences in Chinese and Indian land degradation experience have more to do with the state's role and historical management of different types of land. Specifically, forests were owned by the government in India historically, leading to low levels of degradation despite low income. In contrast, forest land in China was privately/locally owned historically until the 1980s, which led to high levels of degradation. This demonstrates that trends based on the UNCCD data need to be taken as indicative, and there have to be detailed analyses of the pathways which connect socioeconomic variables and land degradation/restoration. We attempt such an analysis briefly in the next two chapters, and a detailed analysis of ten countries is given in a forthcoming book.

Annexure I: List of Countries

The names of the countries alongside the three-letter codes are provided below:

Name	Code	Name	Code	Name	Code	Name	Code
Algeria	DZA	Estonia	EST	Malawi	MWI	Senegal	SEN
Antigua and Barbuda	ATG	Eswatini	ASW	Malaysia	MYS	Serbia	SRB
Argentina	ARG	Ethiopia	ETH	Mali	MLI	Sierra Leone	SLE
Armenia	ARM	Finland	FIN	Mauritania	MRT	Slovak Republic	SVK
Australia	AUS	Gabon	GAB	Mauritius	MUS	Slovenia	SVN
Azerbaijan	AZE	Gambia	GAM	Mexico	MEX	Somalia	SOM
Bangladesh	BGD	Georgia	GEO	Mongolia	MNG	South Africa	ZAF
Belarus	BLR	Ghana	GHA	Montenegro	MNE	South Sudan	SSU
Belize	BLZ	Greece	GRC	Morocco	MAR	Spain	ESP
Benin	BEN	Guinea	GIN	Namibia	NAM	Sri Lanka	LKA
Bhutan	BTN	Guinea-Bissau	GNB	Nepal	NPL	State of Palestine	PAL
Bolivia	BOL	Guyana	GUY	New Zealand	NZL	Sudan	SDN
Bosnia and Herzegovina	BIH	Haiti	HTI	Niger	NER	Suriname	SUR
Botswana	BWA	Hungary	HUN	Nigeria	NGA	Syrian Arab Republic	SYR
Bulgaria	BGR	India	IND	Oman	OMN	Thailand	THA
Burkina Faso	BFA	Indonesia	IDN	Pakistan	PAK	Togo	TGO
Burundi	BDI	Iraq	IRQ	Palau	PLW	Trinidad and Tobago	TTO
Cambodia	KHM	Italy	ITA	Panama	PAN	Tunisia	TUN
Cameroon	CMR	Jamaica	JAM	Paraguay	PRY	Turkey	TUR
Canada	CAN	Jordan	JOR	Peru	PER	Turkmenistan	TKM
Chile	CHL	Kazakhstan	KAZ	Philippines	PHL	Uganda	UGA
China	CHN	Kuwait	KWT	Poland	POL	Ukraine	UKR
Colombia	COL	Kyrgyz Republic	KGZ	Portugal	PRT	United Arab Emirates	ARE
Costa Rica	CRI	Latvia	LVA	Qatar	QAT	Uzbekistan	UZB
Cote d'Ivoire	CIV	Lebanon	LBN	Moldova	MDA	Venezuela (Bolivarian republic of)	VEN
Czechia	CZE	Lesotho	LSO	Rwanda	RWA	Viet Nam	VNM
Dominica	DMA	Liberia	LBR	St. Lucia	LCA	Yemen	YEM

(continued)

(continued)

Name	Code	Name	Code	Name	Code	Name	Code
Dominican republic	DOM	Libya	LBY	St. Vincent and the Grenadines	VCT	Zimbabwe	ZWE
Ecuador	ECU	Lithuania	LTU	Sao Tome and Principe	STP	Brazil	BRA
Equatorial Guinea	GNQ	Madagascar	MDG	Saudi Arabia	SAU	Democratic republic of Congo	CON
Eritrea	ERT					Tanzania	TZA

Annexure II: Measurement of Land Degradation by UNCCD

The United Nations Convention to Combat Desertification (UNCCD) calculates land degradation levels primarily using the UNCCD's Good Practice Guidance (GPG) for Sustainable Development Goal indicator 15.3.1, which is defined as the proportion of land that is degraded over total land area. The UNCCD assesses land degradation using a three-indicator approach, based on:

- Land Cover Change- Land Cover sub-indicator reports degradation in land cover change based on a national assessment of the positive or negative aspects of transitions from one land cover type (e.g. forest, cropland, grassland) to another. Using time series satellite products (e.g. MODIS, ESA CCI), transitions are scored with a land cover transition matrix that distinguishes degrading vs. improving transitions.
- Land Productivity- Based on Normalised Difference Vegetation Index (NDVI) time series, the Land Productivity sub-indicator reports productivity degradation using Earth Observation (EO) data to monitor changes in Net Primary Productivity (NPP) of vegetation. This assessment is based on three metrics, calculated from a time series of annual NPP observations, that are designed to identify changes in the trajectory and level of productivity. Using tools like Trends. Earth (using MODIS NDVI), or ESA's Land Productivity Analysis (LPA), productivity trends are classified into stable/increasing productivity, declining productivity and early signs of decline. Declining areas suggest degradation.
- Soil Organic Carbon- SOC change is estimated using models and baseline maps (e.g. from ISRIC SoilGrids, HWSD). Similar to the above indicators, changes over time in SOC stocks indicate degradation (loss) or improvement (gain).

These indicators are assessed spatially at the pixel level, typically using remote sensing data with 250 m or 30 m resolution, aggregated at national or sub-national levels. The SDG indicator 15.3.1 is calculated by evaluating changes in the sub-indicators over time to determine the extent of land that is degraded over total land area. The extent of degradation measured in each of the reporting periods is compared to the extent of degradation measured in the baseline period to identify whether this area is increasing or decreasing over time. The indicator is calculated by integrating

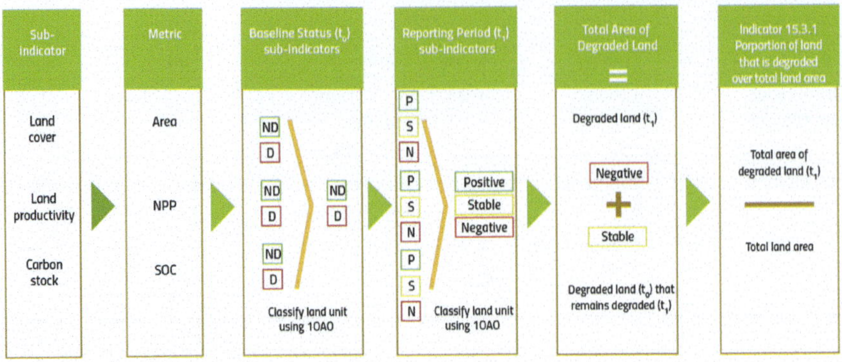

Fig. 3.9 Method of computation of SDG indicator 15.3.1 More information on the construction of the indicator is available on

the sub-indicators using a one-out-all-out (1OAO) method, in which a significant reduction or negative change in any one of the three sub-indicators is considered to comprise land degradation.

The UNCCD reporting template includes the indicator and sub-indicators. The methodology for the indicator is universal, allowing countries to select the most appropriate datasets and national methods for estimating the sub-indicators and determine the most suitable pathway for deriving the indicator. In this regard, the GPG helps countries access and interpret a wide range of data sources for the sub-indicators, whilst ensuring national ownership. At the national level, the institutions that specialise in environmental data are able to work synergistically with National Statistics Offices (NSOs). Thus, the use of the UNCCD's national reports provides a practical and harmonised approach by which countries can report on the indicator.

In practical terms, for the purposes of calculating the indicator, tracking change in the extent of degraded land is a three-step process:

1. Calculate the extent of degradation in the baseline period (t_0) from 1 January 2000 to 31 December 2015. This sets the benchmark against which progress towards achieving SDG Target 15.3 is assessed in each of the reporting periods.
2. Calculate the extent of degradation in the reporting period (t_n), which indicates the recent extent of degraded land.
3. Calculate the change in extent of degradation between the baseline and reporting periods.

The method of computation for SDG indicator 15.3.1 is illustrated in Annexure Fig. 3.9.

Chapter 4
Enabling and Disabling Socioeconomic Factors of Land Restoration: Cases of India and South Africa

This chapter discusses how socioeconomic factors may enable or disable land restoration in two countries, namely India and South Africa.

India

In the data analysis reported in Chap. 3, we could see that India is a country that started with a low land degradation but is currently experiencing a higher level of degradation. India is the seventh largest country in the world having a total geographic area of 328.73 million hectares and occupying only 2.4% of the world's geographical area. However, it supports over 18% of the world's human population and over 18% of the world's cattle population. India is one of the top ten forested countries of the world. Around 69.6% of the total geographical area is categorised as dry lands in the country. According to various estimates, about 29–35% of the total geographical area of India is degraded.[1] According to UNCCD, as of 2019, 9.45% of the land is degraded in the country rising from 4.42% in 2015.[2] A map displaying the reported degraded areas in the selected country for 2019 is presented in Fig. 4.1. Purple shaded areas are identified as degraded areas whereas the yellow shaded areas are the non-degraded parts of the selected country.

One factor that led to a relatively lower land degradation in the past is that related to forests. A major part of forests came under public ownership during the colonial time. This policy was continued by post-independent governments also. Moreover forests that were under the control of princely rulers (or landlords) also come under public ownership during the post-independent period. Though these forests were managed with the goal of timber extraction until 1970s, it changed afterwards. The growing concern about natural environment enabled the management of forests to

[1] *Source* https://www.moef.gov.in/uploads/2023/07/NAP%20final-2023.pdf.

[2] *Source* https://data.unccd.int/land-degradation?grouping=SDG&country=IND.

V. Santhakumar et al., *Land Restoration: A Rational, Grounded and Pragmatic Approach*, https://doi.org/10.1007/978-3-032-26652-1_4

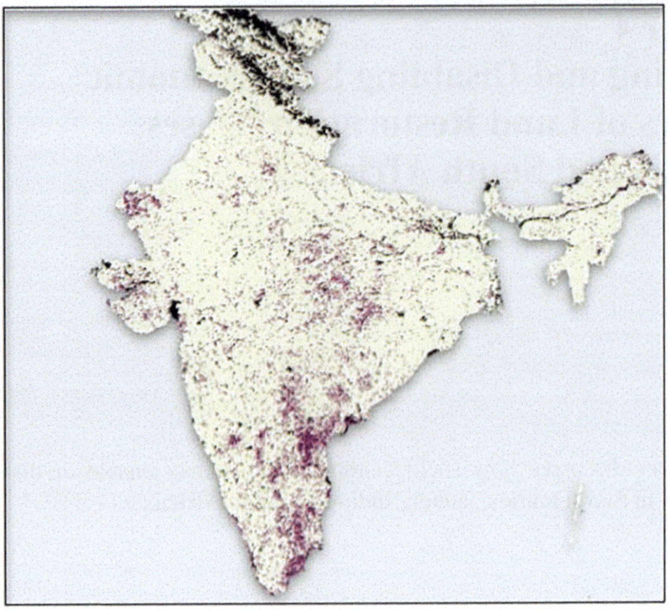

Fig. 4.1 India map presenting land degradation as of 2019. *Source* UNCCD

focus more on conservation goals. Global and national policies and incentives are aimed at such a conservation currently. All these have enabled a relatively higher level of conservation of forests in the past and also recently though there can be issues related to quality and impact on human beings (say through human-wildlife conflicts).

There could be a relatively lower degradation in agricultural lands in the past but that may be increasing currently (and this may be contributing to a higher increase in land degradation at the aggregate level in the country). There was a highly unequal distribution of land in India in the past. A significant share of the population did not have access to land in the country. Landlessness, even in rural areas, was higher then. Though post-independent governments talked about land reforms, the extent of land that has reached the landless people was limited through this route. Though there could have been a redistribution of land through markets, its extent was also limited in India. For example, those who lose interest in cultivation (as they move into non-agricultural professions) can lease out land to those who can use their labour for farming. However such a transfer of land did not happen at scale due to the legal ambiguities regarding tenancy. (Indians did not have good memories of tenancy that existed during colonial period due to the exploitation by landlords, and this compelled many state governments to control or even abolish tenancy.) Those who don't want to cultivate land could have sold it but the extent of this market-based transfer was also not high since most of those who want to expand cultivation may not have access to capital to buy additional land.

Those who move out of agriculture may keep the land by thinking that its price will go up in the market. Land has become a speculative asset, and the relative underdevelopment of financial markets encourage many to invest their money in land. This has increased the price of land, which has become unrelated to the potential agricultural income from it. This too has worked against the transfer of land to those who want to cultivate it. All these created a situation where agriculture did not expand or intensify to the extent that would have happened by considering the population of India. This may have kept land degradation at a lower level in the past.

However the situation has changed during the last few decades. The economic growth has increased food consumption and yielded certain benefits to agriculture. Commercial agriculture has developed. Though the lower-caste groups could not have a notable increase in their farm holdings, middle-castes could expand cultivation. There was also an expansion of animal husbandry due to the increase in consumption of diary and meat products over time. Given the small-holdings of the majority of farmers and their lower incomes, governments have extended a variety of subsidies. These included support prices for crops but also subsidies for irrigation (including that for electricity for pumping water from tube wells), chemical fertilisers and pesticides.

Though there are regulations to avoid the over-use of these inputs, the cost of their enforcement is high since there are millions of small farmers across the country. Due to the need to use the limitedly available public resources for multiple purposes that can be devoted for the enforcement of such regulations is limited. Moreover successive governments are unable to reform subsidies since farmers constitute a vulnerable but politically important constituency. There is an increase in the over-extraction of groundwater leading to the reduction of water table and salinisation, and the excess fertilisers reach water bodies leading to their eutrophication. All this lead to one or other kind of land degradation. Hence land degradation related to agriculture has increased in India over time.

India experienced notable economic growth from 1980s onwards. This has enhanced the demand for minerals both for domestic use and for exports. This period also witnessed privatisation of different industrial activities including mining. Though regulations are in place to see that land is rehabilitated after its use for mines, their enforcement is weak. There is also an indication that the enforcement of environmental regulations in India depends more on the demand from below (through public interest litigations and other strategies), and this could be muted in parts of India where the interest, awareness and capacity of people to 'demand' such enforcement are weaker. This can also be a reflection of the underdevelopment that persists in the country. There are government organisations which collect compensation from organisations which use land for industrial purposes for reforestation and land rehabilitation, but these financial resources are also not used effectively for the purpose. All these have enhanced land degradation due to mining and related industrial activities, though it is concentrated in parts of India.

Though the level of urbanisation is lower in India compared to develop and other emerging economies, it is steadily increasing. Currently nearly 40 per cent of the population live in areas which are designated as urban. This is partly facilitated by the growth of service sector which is happening in cities. It is also attracting a notable section of poorer population from rural areas, who do not have opportunities to earn an income outside low-productive agriculture. This has led to an unplanned growth of cities without adequate infrastructure and public services. Settlement growth can also be seen in rural areas, again without much facilities for the proper disposal of solid and liquid waste including sewerage. These have led to the degradation of land in and around human settlements in both urban and rural areas.

In one sense, India's land degradation in the past and current period is connected to its path of economic development. India was underdeveloped and the resources (mainly land) were not accessible to the majority who were poor in the past. This was not only in the case of agricultural land but also forests. (There was no major political mobilisation to make these resources accessible to a wider section of society, as in the case of China). However, the current increase in land degradation is an outcome of the process of an economic growth that the country has started experiencing from 1980s, but the one that is yet to increase the per-capita income of the country to a threshold where adequate investments are made to conserve land and natural resources. The country could be at the initial phase of an environmental Kuznet's curve in this regard.

The unplanned urbanisation can also be part of this trend—despite urbanisation, the state does not have resources and institutional capacity to moderate its impact on environment and land. The specific nature of its development pattern - with a stagnant manufacturing and growing service sector—may have both positive and negative impacts. Persistence of a nearly half of the population in agriculture may have enhanced the degradation of agricultural land. The land degradation due to industrialisation could be lesser than what it would have been the case, if the country also experienced higher industrial/manufacturing development. (It may be noted that though cities like Delhi face higher pollution it is more due to agricultural activities like burning of crop-waste or emissions from motor vehicles or dust particles due to construction and other activities and not much due to the emission from factories. On the other hand the latter was the major source of pollution in Chinese cities.)

South Africa

In our data analysis we found South Africa started with high land degradation and continues to experience a higher level of degradation. Land degradation is notably higher in South Africa. According to the National Action Plan (2022), approximately 91 per cent of South Africa's landscape is drylands and that makes it susceptible to desertification.[3] Nearly 30 per cent of its land is degraded according to the UNCCD.[4]

[3] *Source* https://www.dffe.gov.za/sites/default/files/docs/nap_desertification_land_degradation_droughteffects.pdf.

[4] Up to 29% of land or 35.6 million ha was degraded till 2019 and most areas are either hit by severe or moderate drought (UNCCD).

Table 4.1 Changing land use in South Africa

Changing land use/cover category	Net area change (2000–2010) Sq. km	SOC 2000 (ton/ha)	SOC 2010 (ton/ha)	2000 total (ton)	2010 total (ton)	2000–2010 loss (ton)
Forest to cropland	1043	61.7	48.6	6438042	5070492	–1367550
Forest to shrubs, grasslands and sparsely vegetated areas	871	46.4	46.4	4045986	4045986	0
Total	1914			10484028	9116478	–1367550
Percent loss total SOC stock (country)						–0.03

Source UNCCD

Notes

(*) sq. km stands for square kilometre or km^2. To convert sq. km to hectares (ha) ×100.

(**) Values for Net LPD and SOC are only for areas where Land Use/Cover is unchanged from 2000–2010.

(***) 'No Data' includes snow, ice, desert areas, water bodies and missing pixels.

(****) Change in SOC due to changing Land Use/Cover derived from IPCC Good Practice Guidance for LULUCF (2006).

(*****) The areas corresponding to marine and other major international water bodies are excluded as out of LDN TSP scope which concerns degradation on terrestrial ecosystems only. Wetlands and smaller sweet water bodies are included as they are an integral part of the surrounding terrestrial areas that deliver the corresponding ecosystem services.

A major part of the country's land is dominated by woodlands, bushland, and grasslands.[5] The most important land-use changes (between 2000 and 2010) are from forests to cultivated land (which occurred in around 1,000 square kilometres) and from forests to bushes and sparsely vegetated areas (in around 870 square kilometres).[6] Both these indicate a transition from 'natural forests' (which may be shaped by the long-standing climatic conditions of the country) to one where intensive cultivation or grazing bring about land degradation (Table 4.1).

[5] Savannas (woodlands and bushlands) and grasslands cover 25.70% and 19.92% of South Africa, respectively. About half of the country is categorised as arid or semi-arid.

[6] Forest cover declined by 1.8% between 2000 and 2010. In contrast, shrubs, grasslands and sparsely vegetated areas increased by 0.1% and croplands by 0.4% during the same period. Between 2000 and 2010, 16.2% of forest showed declining or stressed LPD while 82.8% had either stable or increasing LPD. In sharp contrast, 35.9% of shrubs, grasslands and sparsely vegetated areas class showed declined or stressed net LPD while 63.0% had either stable or increasing net LPD. For cropland, 17.0% showed declining or stressed net LPD while 82.7% showed either stable or increasing net LPD.

In fact, according to the Department of Forestry, Fisheries and Environment of South Africa, similar patterns prevail even for a longer time period between 1990 and 2022.[7] In contrast, the reverse conversion from cropland and shrubs and grasslands to forests during the same period has been miniscule. The National Action Plan of South Africa (2022) discusses that 80% of the land in South Africa is used for agriculture and subsistence livelihoods. About six million people in South Africa depend on agriculture for their livelihoods with the smallholder agricultural sector providing employment for 1.3 million households.[8]

Another aspect that may be relevant to South Africa is communal land. According to Mani et al. (2021), before apartheid ended in 1994, nearly 3.5 million African people were resettled in South African 'homelands', now called the communal areas (Fox and Rowntree, 2001), where enforced high densities of people and livestock resulted in rangeland degradation (Hoffman and Todd, 2000; Meadows and Hoffman, 2002). Both the communal lands historically occupied by the Black farmers and the commercial lands owned by the White farmers currently face the issue of land degradation due to woody plant encroachment. Furthermore, in South Africa, mining is a major source of income. However, mines damage the soil and underlying structure of the land. Different chemicals that are used in mining pollute soil and water courses causing the degradation of land.[9]

South Africa is one among the ten countries for which we analyse the socio-economic enablers or disablers of land restoration. There are interesting patterns among these countries. All countries which have a PCI which is above the median value have either a lesser than median value of land degradation or a decline in degradation over time (if these start with a higher level of degradation). On the other hand, all those countries with a lesser PCI than the median value, have either a higher level of land degradation or an increase in degradation over time if they start with a lower level. However, South Africa is an outlier in this regard. Despite having a PCI that is higher than the median value, it is experiencing an increase in land degradation over time (even though it started at a lower level). A probable reason for this somewhat peculiar situation can be sensed without much difficulty. Though South Africa has a PCI that is higher than the median value, its human development index is below the median when countries are arranged on the basis of this indicator. There is no other country among this sample of ten countries which have a higher than median PCI and lesser than median HDI.

---

[7] *Source* https://egis.environment.gov.za/sa_national_land_cover_datasets.

[8] *Source* https://www.dffe.gov.za/sites/default/files/docs/nap_desertification_land_degradation_droughteffects.pdf.

[9] BFAP (2012) outlines that coal mining in Mpumalanga was calculated and estimated that 46.4% of the total high potential arable soils are found in that province and that 12% of South Africa's total high potential arable land will be transformed. It was also outlined that a total of 326000 hectares of cultivated land will be taken up by current mines and further 439 000ha which has prospecting taking place on them. Even though there are precise execution of well-defined rehabilitation procedures that are done, a degradation from pre mining to post mining land capability is unavoidable. After some years of rehabilitation, the grass cover often deteriorates. Also due to soils being moved too wet, companies cannot be reversed even with deep ripping (BFAP 2012).

The reason for this peculiar situation of South Africa is not unknown. Its Black population that constitute nearly 80% of the population faced the severity of apartheid policies which excluded them from the benefits of the growing economy in the past. This had an impact on their historical acquisition of assets and human capital. The majority among them were poor when apartheid ended. Though post-apartheid governments took steps to address their underdevelopment, the impact of apartheid continues, and a major section of this population is without enough assets and human capital and hence could not gain much from the post-apartheid economic growth of the country (the average growth of the economy during the period has not been remarkable, following a declining trend in the last decade.) Hence the beneficiaries of the economic growth in the country continue to be a relatively smaller section of South African society (which may include the Whites and a small share of the Black population). The growth of PCI of the country has not translated into a significant improvement of human development indicators of the majority.

How does this situation affect land degradation or the possibilities of land restoration? One part of the agriculture is highly commercialised and takes place in very large farms; however, side by side, more than two million smaller farms exist. Even though the soil of the country is not the best for agriculture, those who depend on these small farms have to earn their livelihood from farming or cattle grazing. The educational achievements of the majority (Blacks) are not good enough to get formal sector jobs (salaried employment) in manufacturing or service sector,[10] and hence they have to depend on less productive sectors, and small-scale agriculture is one among them. This led to a situation where small holders may not make enough investments to mitigate land degradation, and it is also difficult for the state to insist on such investments since these people do not have viable alternatives.

Those people who cannot get salaried jobs in manufacturing or services may continue with grazing and agriculture (in not so suitable soils) and this can lead to impacts like woody encroachment. The cessation of the apartheid and the related process of democratisation have encouraged the successive Governments of South Africa to try to meet the basic needs of the majority of the population (Blacks). There was an attempt to distribute small parcels of lands to them as a way to meet their livelihoods. This could have led to the use of a set of large farms and public lands for agriculture by small holders and this can also lead to land degradation.

The persisting problems of poverty and underdevelopment of the majority can also mute their voice when private industries cause land degradation (say in mining areas). This can be noted in contexts where poorer communities interact with private industries which are involved in activities such as mining. The spatial exclusion of Blacks during the apartheid and their underdevelopment have also caused the migration of a section of poorer people to cities. Given their low levels of educational achievements, they may not get jobs in the formal sector, and hence they would become part of the urban poor, and the increase in the share of them would enhance

[10] Though educational progress has been achieved by Blacks, since they form the biggest category of people without schooling and proportion of Blacks achieving tertiary education is still remarkably low. Statistics Government of South Africa https://www.statssa.gov.za/?p=17134#.

the informal settlements in the core and periphery of urban areas. This can lead to land degradation in urban areas also. The migration of people from rural areas to cities and other kinds of socio-economic changes can weaken community norms, and hence it can have a negative impact on the conservation of land and other natural resources through these norms.

The socio-economic transition of South Africa that we have mentioned had its impact on governance as well. Post-apartheid governments are compelled to use a greater part of public resources to meet the development needs (the provision of food, education, housing, electricity, etc.,) of the majority, and a visible improvement on these counts may take time, given the severe underdevelopment of this section of the population when the apartheid ended. Such a use of resources would reduce the availability of public resources for other needs, and it can reduce the ability of the state to enforce its own regulations. This can enhance the land degradation due to commercial enterprises whether these are in agriculture or industry. Hence there can be land degradation due to commercial farming and also different kinds of private industry. Given the importance of mining in South Africa, any laxity in regulating this activity can have a notable impact on land and ecosystems. Due to the peculiar political transition of the country, the same party continues in power after the end of apartheid. These can have negative impacts on governance and the enforcement of regulations which are meant to control land and ecosystem degradation.[11]

[11] Zondo Commission Report https://pari.org.za/wp-content/uploads/2022/09/PARI-Summary-The-Zondo-Commission-A-bite-sized-summary-v360.pdf; https://www.pprotect.org/?q=content/public-protector-south-africa-releases-reports-investigations-it-concluded-second-quarter#; Chipkin et al., 2018: Shadow State: The Politics of State Capture https://www.jstor.org/stable/10.18772/22018062125.

Chapter 5
Enabling and Disabling Socioeconomic Factors of Land Restoration: Cases of Brazil, Australia and China

In this chapter, we discuss how socioeconomic factors enable or disable land degradation/restoration in three nations, namely Australia, Brazil, and China. These countries have sizeable land areas, and their degradation/restoration has implications for the aggregate global picture in this regard.

Brazil

In the data analysis that we conducted as a prelude to country-level analysis (reported in Chap. 3), Brazil is found to be a country that started with a lower (than median) land degradation but is currently experiencing a higher level of degradation. Brazil is the largest country in South America with a total geographical area of 851.4 million hectares. It encompasses 5.44% of the world's geographical area, supporting 2.6% of the world's total population.[1] Further, Brazil harbours the second-largest cattle herd globally, comprising approximately 22.2% of the world's cattle population.[2] It is also the second most forested country in the world, with 62% of the Amazon forests located here.[3] It has an extensive coastline to the east. In addition to harbouring over a third of the Earth's tropical forests, Brazil is home to an extremely rich flora and fauna and a rich diversity of ecosystems including, but not limited to, the Amazon forest.[4]

---

[1] *Source* https://www.worldometers.info/world-population/brazil-population/#:~:text=Brazil%202025%20population%20is%20estimated,(66%20people%20per%20mi%C2%B2).

[2] *Source* https://www.nature.org/content/dam/tnc/nature/en/documents/English.Bain.TNC.pdf.

[3] *Source* https://www.statista.com/statistics/238893/ten-countries-with-most-forest-area.

[4] It also includes the Cerrado (central plateaus, covering 21% land area), the Atlantic Forest (forests which extend along the Atlantic coastline), the Caatinga (desert shrubland in the northeast).and the Pantanal wetlands (encompasses the world's largest wetland area, located along the western border).

V. Santhakumar et al., *Land Restoration: A Rational, Grounded and Pragmatic Approach*, https://doi.org/10.1007/978-3-032-26652-1_5

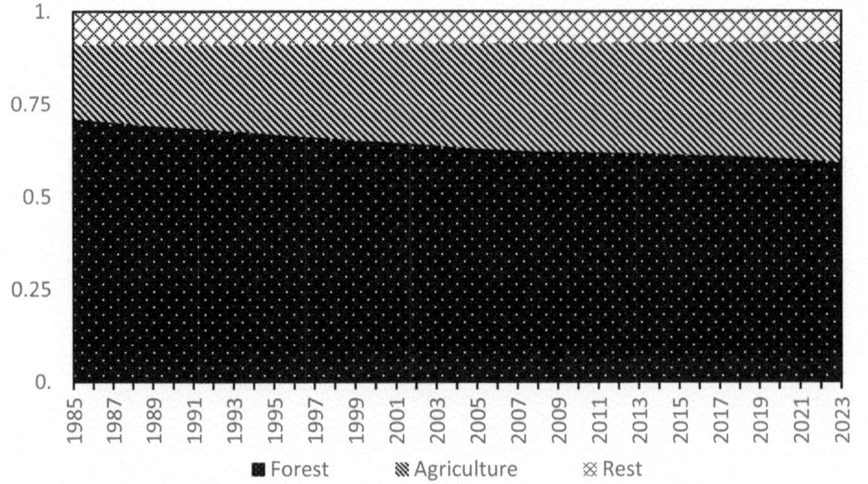

Fig. 5.1 Land use pattern overtime. *Source* MapBiomas

According to data from MapBiomas, Brazil has lost 87.2 million hectares of native vegetation between 1985 and 2019.[5] More recently, as of June 2024, Brazil's degraded land surface has increased to 140 million hectares, according to the latest information from the Department of Forestry of the Ministry of Environment, and deforestation is expected to continue increasing at a considerable rate.[6] We focus on how land use pattern has changed over time in Fig. 5.1.

As observed from the above figure, the area under forest has shrunk from 71% in 1985 to 59% in 2023. Concomitantly, the area used for agricultural, pasture and forestry purposes has increased from 19% to 35%. Other land uses, that include shrubs and herbaceous vegetation, water and non-vegetation land (that includes urbanised areas, waters, industrial areas and mining among others) have remained fairly stable.[7] However, within the non-vegetation land category, the land used for mining purposes and as pasture lands have increased by 460% and 80% respectively during the same time period.

[5] Since we do not have UNCCD country report for Brazil, we rely on other estimates. The MapBiomas project is an initiative of the Climate Observatory co-created and developed by a multi-institutional network involving universities, NGOs, and technology companies with the purpose of annually map Brazil's land use and land cover and monitor the changes in the territory.

Native vegetation refers to plant life that naturally exists in a specific region without human introduction or modification. It encompasses all plant species, communities, and associations indigenous to that area, meaning they have evolved and occurred there naturally. Source for the information on the degradation levels- https://brasil.mapbiomas.org/en/2020/08/28/o-brasil-perdeu-area-de-vegetacao-nativa-equivalente-a-10-do-territorio-nacional-entre-1985-e-2019.

[6] *Source* https://initiative20x20.org/restoration-projects/revitalizing-landscapes-damaged-intensive-agriculture-brazi.

[7] The graph is based upon authors' own calculations with data from the mentioned source.

Since 60% of Brazil's total land area is covered by the Amazon forest and Brazil's economy experienced rapid transformation in the last fifty to seventy years, land degradation is relatively recent in Brazil in historical terms. Olivieria et al. (2022) discuss that in the 1960s, only 2% of Brazilian farmers had agricultural machinery. In the 1970s, Brazil experienced rapid urbanisation and middle-class expansion, increasing the purchasing power of households and changing food consumption patterns. Domestic food production was insufficient due to poor agricultural technology and land management, while food imports were becoming unsustainable.

The land used for agriculture increased by 5% between 1985 and 2023—from a meagre 2% to 7%. At the same time, land used for pasture management doubled during the same period. The share of the urban population also increased by 40% since the 1960s. These broad patterns emphasise how land is viewed as a critical input for all major economic activities in Brazil. Further, as Brazil experienced significant economic growth since the 1960s, it created additional pressure on land.

There are two salient features about land distribution in Brazil—extreme land inequality and inadequate enforcement of land use laws. As mentioned earlier, 1% of the landowners control 45% of all agricultural land—who are major contributors of land degradation. The Brazilian government has undertaken various initiatives to address land inequality, a persistent issue rooted in the country's colonial history. National Institute for Colonisation and Agrarian Reform was established to implement land reform and has been instrumental in redistributing land. Between 1988 and 2013, it facilitated the transfer of nearly 77 million hectares of land to landless families (Albertus et al. 2018). In a recent judgement, to improve land management sustainably, Brazil's Supreme Court mandated the federal government to seize lands from owners found guilty of illegal deforestation or causing wildfires. This ruling also prohibits the regularisation of such lands, aiming to deter environmental destruction and discourage land grabbing.[8]

Despite these efforts, there exist considerable political resistance from large landowners and inadequate enforcement and implementation of land redistribution. As a complementary approach to address these issues, landless or small farmers resort to market mechanisms to buy land from willing sellers. Under the market-assisted land reform (MALR) programme, the government supports and facilitates such market transactions. While this channel is politically less problematic, critics argue that it's mostly the transfer of low-quality land (Fitz, 2018). Thus, land inequality continues to persist.

Governance in Brazil has significantly impacted land management, contributing to degradation. It is the weak enforcement of regulations that contributes to low implementation of mandated restoration on private land, exploitative mining with low accountability, and illegal land use. A substantial portion of Brazil's land, especially in the Amazon, lacks clear legal ownership. This ambiguity fosters illegal land grabbing, deforestation, and speculation. For instance, between 2019 and 2021, 44%

---

[8] *Source* https://apnews.com/article/brazil-amazon-pantanal-deforestation-supreme-court-3cec1105fefe6a33e097e6c5f575c739.

of deforestation in Amazonia occurred on private lands, with an additional 18% on undesignated public lands (Duchelle et al. 2018).

Brazil's rapid economic growth and structural transformation over the past few decades have not been in tandem with ecological restoration. As discussed previously, Brazil has transitioned from agriculture to services—however, the reduced reliance on agriculture has not eased pressure on land. Global competitiveness in agricultural commodities and rapid urbanisation due to the shift towards the service sector have only added to pressures on land, leading to degradation. Finally, successive Brazilian governments have exhibited markedly different approaches to land degradation, particularly concerning deforestation and environmental governance. Left-leaning governments focus on social justice, agrarian reform and indigenous rights-making attempts towards legalisation of informal tenure. In contrast, right-leaning governments focus on agribusiness, privatisation of land and rollback of land reform. While recent efforts signal a renewed commitment to combating land degradation by the current government,[9] the effectiveness of these measures will depend on sustained political will and enforcement.

Australia

Australia is a country that has witnessed a relatively low level of land degradation; further, the rate of degradation has also been slow. Australia is the world's sixth-largest country and the driest inhabited continent. It has a land area of approximately 7.7 million square kilometres and a population of around 27 million.[10] (Reserve Bank of Australia, 2025) Even though the average population density of the country is only 3 per square kilometre, the areas close to the coasts are population dense as they comprise as high as 87% of its population.[11] (SoE 2021) Australia has diverse ecosystems ranging from the tropics in the north to the arid interior, alpines in the southeast, and temperate regions in the south and is one of the twelve most biologically diverse nations in the world, the only developed nation to have this 'megadiverse' status. Australia is 70% arid or semi-arid.[12] For the year 2019, 8.19% of the total land area (62.86 million hectares) was degraded. It was reported 7.3% for the year 2015.[13] (UNCCD, 2022a).

Australia has specific challenges and opportunities in terms of land restoration. The first set of challenges comes from tough biophysical conditions in the form of 50% of its landmass being deserts, while 70% of the country experiences arid and semi-arid conditions that can lead to land degradation even without much human

[9] *Source* https://www.reuters.com/sustainability/climate-energy/brazils-government-aims-raise-2-billion-financing-with-new-eco-invest-auction-2025-04-28

[10] Reserve Bank.

[11] DCCEEW. State of the Environment Report 2021.

[12] DCCEEW. Outback Australia - the rangelands.

[13] UNCCD Report, 2022.

action. These can be aggravated by climate change (which is an outcome of human actions all over the world).

The second set of challenges may come from viewing this land mass and associated resources as an important input to economic prosperity by society. This is somewhat inevitable in any country/economy which may try to maximise its economic output with its own resources, and when surplus land is available, it will be treated as such an input. This has led to development of large commercial farms and cattle ranches and exploration and extraction of minerals to meet domestic and, more significantly, an ever-increasing global demand. Commercial agriculture and mining can therefore be major causes of land degradation. A case in point is that 6.1 million hectares of primary forest were converted to other land uses including agriculture, mining, urban expansion and forestry since 1990. To add to the concern, clearing and re-clearing of native vegetation is taking place without proper authorisation. However, there are a few enabling factors for land restoration in Australia.

It is a developed country with a higher per capita income as well as higher human development. It is also an industrialised economy in which, though agriculture is an important economic activity, industry and services are relatively more important sectors of the economy. The share of the population which depends on agriculture is very low and is declining. The size and productivity of agricultural farms are high. Small farms wherein people depend on small parcels of land for their livelihood are fewer than other economic units. Being a country with a notably higher level of human development, most Australians have reasonably high levels of education that can get them jobs in industrial and service sectors. This is also evident from the settlement pattern. Most people live in a relatively small part of the country which is highly urbanised. While regulatory gaps may exist and their enforcement is challenged by economic pressures and a rapidly increasing urban population, urban planning and the provision of public services are at a higher level relative to global patterns.

Longer periods of economic and human development, and also the presence of a reasonably functioning democracy and related institutions, can equip Australia to handle issues of land degradation relatively well. There are regulations related to land use in commercial farms and for other activities such as mining, and these are enforced (whereas in many other countries such regulations exist but are not enforced). This is on the back of the availability of public resources resulting in high enforcement capacity of the government and the impact of education and awareness of environmental problems among the population, encouraging them to raise their voice for controlling negative externalities.

These factors contribute to checks and balances on the land degradation front on one hand and effectiveness of land restoration efforts on the other, ensuring a low level of degradation in the past and a low rate of degradation in the present. This does not mean that Australia is not experiencing land degradation. There are challenges for land restoration in the country, and it may require specific steps.

China

Based on our analysis, China started with a higher-than-median land degradation, but had a slower pace of degradation over time. It is an interesting case among these countries. In fact, land degradation was so severe in China that the percentage of area under forests came down to a single digit at the peak of its socialist period (Zhai et al. 2017). This is not surprising since forests were not owned by the national government (unlike India, which could control their use by people). Though the Chinese did not have private property rights during the socialist period, local governments and communes ensured that people had access to land for cultivation to meet their daily needs. This and the growing population pressure led to the increased use of almost all types of land for agriculture and the sharp decline of forests.

However, the higher levels of economic growth after the late 1970s changed the combination of economic activities. This growth was driven primarily by manufacturing development, and through this China could become the top supplier of many manufacturing products in global markets.[14] The youth population who completed school education in the countryside started moving to cities and special economic zones where manufacturing developed (Zhang et al. 2003). This reduced the dependence on agriculture as an occupation and on land in rural areas. This development of manufacturing led to the growth of the service sector also, mainly to meet the needs of the domestic economy which is getting integrated with global markets.

There could be different kinds of environmental changes associated with this economic development of China. Urban environmental problems were aggravated and air pollution was a major problem in its cities (Zheng et al. 2024). This was contributed to by the booming manufacturing development. However, the reduction of employment in agriculture and greater dependence on markets for food grains reduced the compulsion of rural families to use the land intensively. The participation in international trade also enabled China to import food or inputs for food production. China imports a significant quantity of farm products from other countries, including the US, to sustain its meat production (Elobeid et al. 2021). Hence, the increase in domestic consumption of food and related items did not exert severe pressure on its land to produce these or the raw materials (like cattle feed) for this purpose.

There are also other developments. Manufacturing and service sector growth increased the incomes of people. Though there was a spread of basic education during the socialist period, it accelerated after the opening up of the economy. Poverty declined rapidly. The decline of poverty and human development had impacts on other fronts. Population size started to decline. China is a country which experienced rapid urbanisation during the last few decades. Those who move out of rural areas in search of employment are not that interested in going back and building their lives there. Though this has led to a crowding and the need for substantial new construction in cities, the effect of population pressure in rural areas has eased.

---

[14] *Source* https://blog.allegisglobalsolutions.com/what-country-is-the-worlds-next-manufacturing-hub.

Economic growth has increased the quantum of public resources which are available to the government. Its capacity to invest in public goods (including the restoration of ecosystems) has increased. Hence there is a greater willingness on the part of the government of China to conserve the natural environment. This can be seen in actions towards land restoration as well (Li and Pimm 2020). There are projects which finance farmers to dissuade them from cultivating in certain areas so as to restore the land.[15] It has also helped a notable increase in the area under forests to about 20 per cent recently, from a low of around 8 per cent a few decades ago. Hence China's case is one where the quality of the natural environment has declined during the socialist expansion of production and also at the initial phase of rapid economic growth (after opening up its economy), but which could witness an improvement in this regard after crossing a threshold of economic development or attaining a higher level of per-capita income. It could be a typical recent case where the conjuncture of the Environmental Kuznet's Curve seems valid. (It speculates that there could be a decline in the quality of the natural environment at the initial stages of economic development, but there could be a reversal in this regard after crossing a threshold of such development.)

A Comparison of Five Countries

We have seen the situation in China, where there was an increase in land degradation a few decades ago, but it has moderated recently. It is easy to compare other countries with China in this regard. Though India is as populous as China, it did not face as much land degradation in the past, but it has not moderated recently as much as the latter. The development path of India in the past was not one that enhanced the access to land for the majority of its poorer people. Its development after economic reforms has not reached a threshold level similar to that of China to have a moderating impact on land degradation.

Though Brazil and South Africa have relatively higher per-capita incomes, these do not translate adequately into investments and action for a better-quality environment (including those for restoring land). The fact that the benefits of economic development were mainly derived by a section of society, and others were not equal participants of the process of development, was the main reason. To a great extent, the inequalities in terms of economic and social positions in these two countries have prevented them from being able to take adequate steps towards land restoration.

Though Australia has less fertile land and faces challenges due to global climate change, it could ensure that land degradation was lower (than the median) in the past, and it continues to be so currently. The long-standing economic development and policies which ensured that the majority of its people benefit from such a development, combined with reasonably functioning institutions and governance, could ensure this outcome. However, international trade and global climate change pose newer challenges to land in Australia.

---

[15] *Source* https://forestsnews.cifor.org/53502/in-china-paying-farmers-to-restore-forest-landscapes?fnl=en.

Part II
Managing Trade-Offs, Incentives and Disincentives for Land Restoration

Chapter 6
A Systems Approach to Understand the Interplay of Enabling/Disabling Factors for Land Restoration

It is noted that there are numerous, interrelated and complex causes of land degradation and the same causal factor could lead to diverging consequences in different contexts because of its varying interactions with other proximate and underlying causes of land degradation (von Braun et al. 2013). This warrants a systems analysis wherein we integrate socio-economic factors with land degradation patterns. Systems approach is used to understand the linkages between human actions based determinants in local/micro level projects of landscape management (for a review, see Pfeifer et al. 2023). How does socio-economic system's complexity leads to soil degradation is analysed (Gambella et al. 2021). However we have not seen cases where a systems approach is used at the country level to understand the possible connections between socioeconomic factors which enable/disable land degradation. This situation warrants an analysis of enabling and disabling socioeconomic factors and how the interaction of these create divergent pathways in different countries leading to land degradation and/or restoration. This may require a systems perspective which see how different factors interact and create specific outcomes in a particular context. The purpose of this chapter is to develop such a systems perspective[1].

Outlining a National System of Land Restoration

The land in a country can be broadly divided into four categories (Fig. 6.1). This categorisation is based on who owns and controls the land. There can be other categorisations but we use the ownership of land since it may help, to some extent, to get mutually exclusive types. It is also based on a presumption that the restoration of land depends on the incentives (and disincentives) of people/firms which own/

[1] A previous version of the material in this chapter benefitted from the editing and making of figures by Apoorva Bose.

V. Santhakumar et al., *Land Restoration: A Rational, Grounded and Pragmatic Approach*, https://doi.org/10.1007/978-3-032-26652-1_6

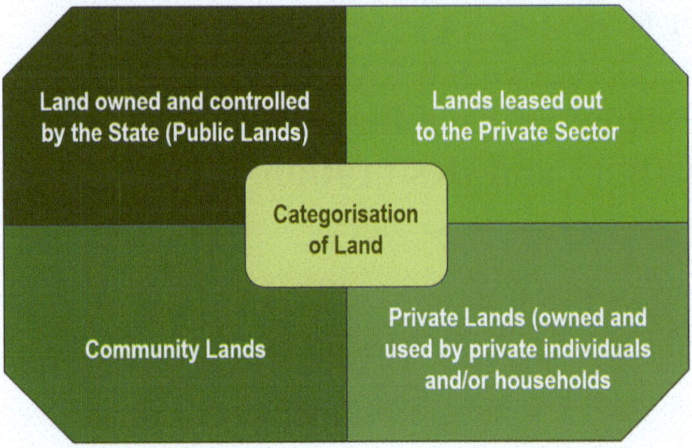

Fig. 6.1 Categorisation of land

control land. Though there are common factors which may determine the degradation or restoration of all these different categories of land, each type may have specific issues, and these are discussed in the following subsections.

Private Land (Land Owned and Used by Individuals and/or Households)

The restoration of privately owned lands depends on the incentives of individuals and households. However, these incentives are shaped by different factors. These include laws governing the use of land, market incentives for land restoration, dependency on land, the impact of state policies, interventions by NGOs etc. These are discussed briefly in the following subsections and compiled in Fig. 6.2.

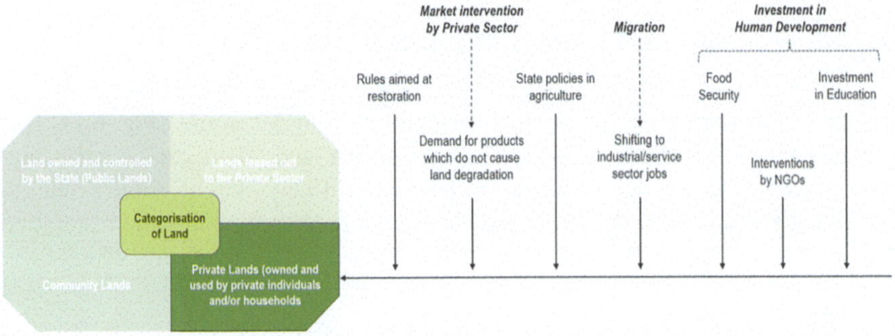

Fig. 6.2 Factors that enable restoration of private land

Laws/Regulations That Demand the Restoration of Private Lands

There may be countries where the government may impose certain restoration or forest-protection practices on private lands. For example, Brazil has such a provision. Its Forest Law requires that rural landowners designate and maintain a percentage of their property under forest cover. This varies from 20 to 80 per cent depending on the type of vegetation and the property's geographical location in the country. It also requires all landowners to restore deforested areas on their properties.[2] The United States uses conservation easements which may limit the use of private land to enable conservation.[3] The landowners who get into such agreements get tax credits. Canada provides incentives for taking conservation measures on private lands.[4] Hence, there can be regulations or direct incentives mandating/encouraging the restoration or protection of forests in private lands in different countries (Ding et al. 2017).

Market Incentives for Enabling Restoration

There are incentives for restoration which are shaped by markets. The increase in the cost of labour (including that which is needed for supervising farm operations) may encourage farmers to reduce the intensive cultivation of lands owned by them. This hike in wages could be due to the increase in the demand for labour in local or national markets. Migration opportunities could be one such reason.[5] There can also be an increase in the price of farm products whose cultivation supports the restoration of the land. Products from trees and perennial plants (as in the case of agro-forestry) may fetch a higher revenue for farmers in certain cases and this may encourage them to move towards land restoration.[6] In certain parts of the world, land which is not productive for agriculture may have opportunities for the development of wildlife tourism and other such activities, which may also enable restoration. Private companies which see restoration as a business can strengthen these incentives for restoration. There can also be for-profit or not-for-profit intermediaries which can facilitate the transfer of carbon finance to private landholders, which too can facilitate restoration.

However, we need to mention two limitations in this regard. Such market incentives may not be adequate in the case of all kinds of land or land use. Moreover, such an enabling role of private companies may work well only if the incentives of

[2] http://climatepolicyinitiative.org/wp-content/uploads/2017/10/FOREST2.pdf.

[3] https://nationalcenter.org/ncppr/2008/05/01/conservation-easements-the-good-the-bad-and-the-ugly-by-dana-joel-gattuso/.

[4] https://www.cbd.int/financial/fiscalenviron/canada-fiscalmeasures.pdf.

[5] Such a connection is observed in Nepal. See Nepal et al. (2022).

[6] This seemed to have happened in the Kerala state of India. See, Mani et al. (2020).

producers (due to state policies and markets, in general) are also aimed at dissuading cultivation practices that result in land degradation.

The Indirect Impact of State Policies

Policies that are not aimed at restoration may have an indirect positive or negative impact on the incentives for restoration in private lands. It is noted that the provision of subsidies for certain kinds of agriculture or inputs may facilitate intensive cultivation with negative impacts on restoration (McFarland et al. 2015). There is scope for redesigning subsidies to create incentives for restoration (Ding et al. 2017). Similarly, state policies can minimise such 'distress' actions of people. For example, the absence of food security may encourage people to cultivate food crops even in those lands which are not appropriate for this purpose, and this may lead to degradation. This can be avoided if governments can ensure food security through market interventions or the provisions of subsidised food grains. Moreover, governments have an important role in providing education and enabling people to move out of agriculture which can also have a positive impact on land restoration.[7] We do not presume that the abandonment of land has no negative impacts on its restoration; policy interventions may be needed to mitigate these negative impacts.[8]

Interventions by Non-governmental Organisations

Different kinds of altruistic or not-for-profit, non-governmental organisations (NGOs) encourage and enable the restoration of private lands. The discussions in the previous paragraphs may help us to identify those factors which may determine the success of these NGOs. In contexts where markets and state support provide incentives to move towards land use which facilitates restoration, the interventions of these NGOs could be an additional enabler. These organisations can solve the information problems by creating an awareness of the need to restore land and the options which are available for such a restoration without affecting the expected private gains. However, in those cases where private landholders do not have the incentive

[7] Non-intensive cultivation or fallowing of land by those who take up non-agricultural jobs (mostly in urban areas) can be a reason for an increase in the area of forests. This has happened in Europe. For example, the European environmental agency notes: Forest area in Europe has increased since 1990 by 17 million hectares (ha) of which more than half are planted forests. This has been the result of afforestation (e.g. planting and seedling of trees on land that was not previously forested) and through natural expansion of forests such as on abandoned land. https://www.eea.europa.eu/soer/2015/europe/forests#:~:text=Key%20trends,such%20as%20on%20abandoned%20land.,such%20as%20on%20abandoned%20land.

[8] https://www.openaccessgovernment.org/land-restoration-recultivated-abandoned-farms-land-cropland-mitigate-carbon-climate-change/136479/.

to adopt restoration-oriented land use (due to inadequate market signals and/or state policies), the task on the part of these NGOs may be difficult. In such cases, it is not surprising if people continue with practices which lead to land degradation in spite of persistent efforts by NGOs.

Community Land

This category includes land that is used by people but is not privately owned. There could be informal or traditional rights which determine access to such land. For most purposes, these may have the features of common property resources (CPR). Though these were part of the CPR, there need not necessarily face what is referred to as the 'tragedy of commons' as noted in Hardin (1968). This could be due to informal norms enforced through communities.[9] (We discuss these in detail in another chapter in the book). However, the folk theorem (Maskin, 1986) demonstrates two conditions for the sustenance of such a collective action (to avoid the tragedy of commons): the infinitely repeated interactions and the sufficiently low discount rates. Poverty may lead to higher discount rates[10] and the breakdown of collective action that is needed to use community lands sustainably. Poverty and vulnerability (and the absence of governmental actions to mitigate these) can enhance the discount rate to people.[11] This would mean that people would be focussing on immediate gains by neglecting the longer-term ones. This can have a negative impact on collective action within even those communities which are closely knit and depend on natural resources. This can be one reason for the degradation of community lands in less developed areas, like sub-Saharan Africa.[12]

This again shows the importance of poverty eradication, food security or even access to healthcare and other such basic needs to minimise these land-degrading practices and to ensure restoration. This problem need not be due to any intrinsic problem of community lands, and we have seen in the previous section that there can be degradation even in privately owned lands if people do not have the incentives to practice agriculture (or land use) that ensures long-term sustainable management of land and its restoration.

[9] That is key insight of studies by the Nobel Laureate Elinor Ostrom and her associates.

[10] We do not discount the counter-evidence. There are cases where poverty may lead to a lower discount rate. Refer Sarwosri & Musshoff, 2020.

[11] There are several studies which indicate that poor people are inpatient and may use a higher discount rate. For example, see Tanaka et al. (2016); Haushofer and Fehr (2014) and Hansen (1979).

[12] For an understanding of the extent of community lands in Africa, see Alden (2017).

Shorter-Period Interactions Among People May Reduce the Incentives for Collective Action

The second insight from the folk theorem is the importance of infinite (longer-term) interactions in sustaining collective action to protect the CPR including community lands. However, socio-economic changes of different kinds may work against these infinitely long interactions. Sections of people may move out (migrate) and there can be periodical immigration. People may start depending on wider labour, commodity and input markets which may reduce the need for interactions (or interdependence) within communities and the use of land or cultivation may become less important as the source of income. All these may reduce the incentives for people to participate in, and abide by, norms enforced through informal collective action. It can also lead to the degeneration of community institutions which have enabled the sustainable management of land in the past. This factor (along with poverty and vulnerability) may lead to the breakdown of collective action, and the degradation of community lands (including pasture lands) in different parts of the world.[13] When collective action breaks down and that leads to the degradation of land, the government can promote alternative management mechanisms. These may include a variety of forms, like privatisation, enforcement of a quota, co-management or even state control. However, the effectiveness of these possible interventions may depend on the capability and responsiveness of the state, an issue that we take up in a later section.

Role of Non-governmental Organisations (NGOs) in Community Lands

These organisations can play an important role in enabling collective action when sections of people have lost their interest/incentives to be active participants in such an action. Such an enabling role on the part of NGOs may include (a) awareness creation; (b) strengthening the bargaining power of those who depend on the resource; (c) supporting the enforcement of community norms; (b) enhancing the stake in the community resource; and so on. However, NGOs may face challenges in enabling collective action where socio-economic change has significantly reduced the dependence on community lands among a sizeable section of users. There can be ethnic and other divides among people which may be leading to the breakdown of collective action. The NGOs have the potential but may face challenges in these contexts too. The incentives which are connected to the community lands are summarised in Fig. 6.3.

[13] For a description of the situation in a case in Central Asia, see https://news.climate.columbia.edu/2020/07/01/kyrgyzstan-pasture-access-disputes//.

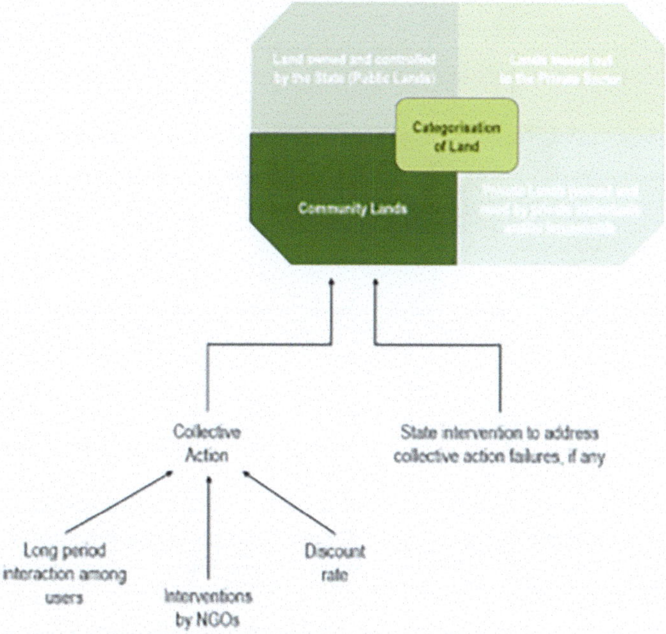

Fig. 6.3 Incentives connected to community land

Land Leased Out to Private Sector

In many countries, governments allocate land to privately or publicly owned corporations and companies which involved in economic activities. These may include timber extraction, development of plantations, mining of minerals, creation of industrial estates, etc. When publicly owned land is allocated for this purpose, these are either uncultivated lands or those which have different levels of vegetative cover, which may be used or cultivated by communities. A certain level of land degradation is inevitable in such lands as part of their use for economic activities. These activities are useful for enhancing social welfare including the creation of jobs (if the society/state ensure these). This may lead to income growth and human development which may enable land restoration in other areas. Hence, we do not take a view that the use of land for this purpose is inherently against the restoration of land.

However, the use of land in such areas can be carried out in ways to ensure that the land degradation that is associated with the economic activity (carried out by the leasing-in company) is not severe. Theoretically, the land use of such land can be through means which do not increase the depletion of nearby waterbodies or do not add to the carbon dioxide in the atmosphere. It can be ensured that only that part of the land which is really needed for economic activity is used, and the remaining is conserved. There are cases like mining where land that is used for this purpose can

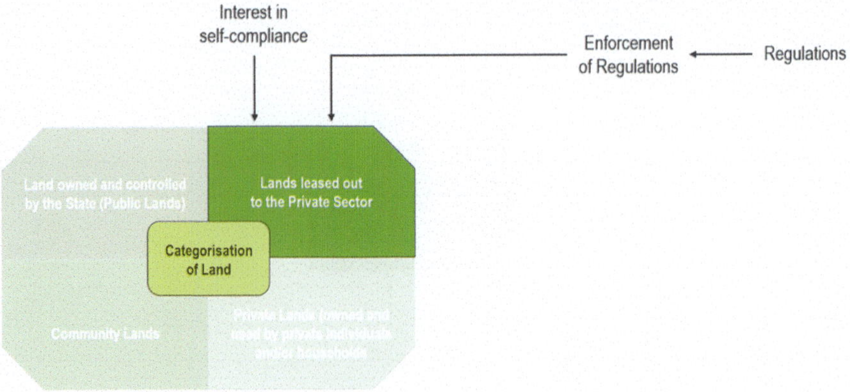

Fig. 6.4 Land leased out to private sector

be restored in the best possible way after the cessation of that economic activity.[14] However, the corporations which get the land from the government may not have the incentives to take these precautionary measures unless these are mandated or regulated by the latter.

Hence, the key issue is the presence of government regulations on the use of these lands, and the effectiveness of the enforcement of these regulations (Fig. 6.4). Even if governments make regulations by following international best practices, these need not be enforced. There are two aspects here: first, there may not be an adequate popular demand for the enforcement of these regulations, and these are connected to underdevelopment, ill-governance, and politics—issues that we take up towards the end of this essay. Secondly, the observance of these regulations cannot depend solely on enforcement by governments. Those who manage the corporations or companies should also feel compelled to adopt these measures voluntarily. Such a voluntary adoption or compliance also may be connected to the level of development—an issue that we discuss towards the end.

Non-governmental organisations (NGOs) may play a role in this regard by creating higher demand by citizens or pressure on governments to formulate and enforce stringent regulations regarding the use of this type of land. The success of such actions also depends on domestic governance and politics. There can be international pressure also for this purpose. National governments may be motivated to take certain actions in this direction due to financial incentives and/or negative publicity that may be created due to international pressure.

---

[14] Technical solutions exist for this purpose. For example, https://semspub.epa.gov/work/HQ/159627.pdf.

Land Owned and Controlled by the State

In almost all countries, a significant share of land, including forests, is owned and controlled by governments. The conservation or management of forests is often carried out by government organisations.

Key Issues Regarding Conservation of Public Land

There can be a stated intention to protect these lands (including forests) in most countries, but the actual protection may be shaped by financial and human resources which are available to governments and the possible challenges. This is linked to the nature of governance in the country. Given that the conservation of these lands is to be carried out usually by public sector organisations, there can be issues of ineffectiveness, inefficiency or corruption, and all these may require vigilance on the part of citizens. Whether citizens are informed and interested to demand a higher level of conservation in a context could be a deciding factor (which is connected to human development—an issue that we take up in the following section). In certain cases, conservation may be achieved at higher economic and social costs. There is an issue of economic efficiency when the potential income from land and forests without compromising conservation goals is not realised. (For example, though nature-based tourism can be promoted without damaging forests, public sector organisations may not have the incentive to maximise incomes from such activities without harming the natural environment[15]).

There can be distributional issues too. The actual effort towards protection (whether there are laws or not) may be determined by the political economy. If users of forests directly or indirectly have electoral clout, it may discourage democratically elected governments from going ahead with stringent protection. On the other hand, there are cases where the actual users of forests may not be politically powerful, and that may encourage the government to neglect their (genuine) issues. This may be true in other lands also which are owned by the government, like rangelands, which are used by pastoralists and communities. Higher social costs arise when forests are conserved by denying the rights of sections of direct users, especially poor and marginalised ones. For example, there is a perception that the conservation of forests in India is achieved by denying the rights of its tribal people, and legal measures which are made to mitigate this problem are not implemented properly.[16] In many contexts, the middle class may take a pro-conservation attitude even if it is not shared by workers and capitalists who gain from natural resource-based industries. The government's position would be shaped by the relative position of these different classes/groups in the political economy. These issues are compiled in Fig. 6.5.

---

[15] This is noted in the literature. For example, Santhakumar (2009).

[16] That is the situation with respect to the implementation of the so-called Forest Rights Act.

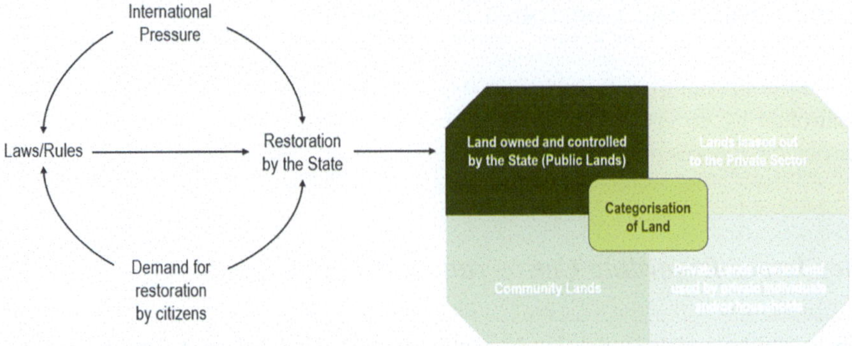

Fig. 6.5 Land owned and controlled by the state

A Set of Meta-Factors

The aggregate outcome in terms of land restoration in a context depends on the interplay of the factors mentioned and a set of meta-aspects which may influence all or most of these factors. There can be two kinds of interplay. There may be a possibility of substitution—which means that even if one enabling factor is absent, the presence of others may facilitate restoration. For example, even if no law mandates (or government policy supports) restoration in private lands, economic factors that encourage people to move out of intensive land use may facilitate restoration. There can be issues of complementarity also wherein the presence of one enabling factor may not make much impact because of the absence of a complementary input. The institution of rules which mandate restoration (as in the case of restoration of mining lands in many developing countries) may not be effective if there is not enough investment for the enforcement of these rules, or there is not enough self-compliance on the part of companies.

The meta-features that impact most, if not all, enabling factors are human development, governance, and political economy. The way political economy impacts the restoration of public lands is discussed in the previous section. Political economy can impact governance and state policies which impact human development (say, through investments in education, supply of basic food to people at affordable prices, etc.).[17] Such a human development may reduce the dependence on land as the main source of subsistence, especially for smallholders. When these people move out to industrial/service sector jobs in urban areas, the disincentives for restoration in private lands may decline in many contexts (especially if there are only limited alternative uses of land in rural areas). This depends on how restoration is viewed. The abandonment of agricultural land can work as passive restoration, improving, for example, water regulation, soil recovery, nutrient cycling and increased biodiversity. Human development may have a positive impact on the popular demand for a better-quality

[17] Santhakumar, V. (2014) The Roots of Ill Governance and Corruption, Sage, New Delhi.

natural environment (which has features of a public good), and that may also enable land restoration.

Political economy has an impact on the conditions with which public land is allocated to corporations for productive activities. In different countries, including the USA or Brazil, the mainstream political parties differ in terms of their approaches towards the privatisation of public lands for economic activities and the restrictions to be imposed on these private entities. In essence, the incentives and disincentives for land restoration in different categories land depends on these meta-factors of human development, governance and political economy.

Chapter 7
Need to Combine Science with Socioeconomic Priorities in Land Restoration

There are different ways to restore land[1]. Allowing land to naturally regenerate is one such strategy. Planting multiple indigenous tree species is another. In addition, there can be plantations of one or more (indigenous and/or exotic or brought in from other localities) species of trees which can generate timber, which if harvested sustainably, can sustain land cover and facilitate land restoration. Such a harvest of timber and other products can generate financial incentives for restoring land.

There are apprehensions, especially among ecologists and environmental activists, about the use of single-species tree plantations for land restoration.[2] It is true that such plantations may not be like secondary and mature forests in terms of multiple environmental benefits. However, is it desirable to neglect plantations as a practical and viable strategy towards land restoration? What are the trade-offs between natural regeneration and plantations as different means towards land restoration? These are the questions that are briefly addressed in this chapter.

Focussing on Three Environmental Benefits of Restoring Land

In this chapter, we focus primarily on three objectives of land restoration. First, is the prevention of land degradation. This may require a sustainable vegetation cover over the land. Such a cover should be adequate to prevent erosion, as well as the decline of soil fertility. The second objective is the protection of biodiversity. It is needless to mention the benefits of conserving pools of species and genetic biodiversity worldwide (Balmford et al. 2002). Even if these are not used currently, there is value in

[1] Santhakumar's discussions with Daniel Piotto, a professor of forest sciences from the University of Bahia, Brazil inform this chapter.

[2] It is accepted that plantations are needed to meet consumer demand and hence, may reduce the pressure on natural habitats (WRI, 2016).

V. Santhakumar et al., *Land Restoration: A Rational, Grounded and Pragmatic Approach*, https://doi.org/10.1007/978-3-032-26652-1_7

conserving them by considering the option value (the value when new information is available in the future) and the irreversibility if this diversity is destroyed. The third objective of land restoration is the enhancement of carbon dioxide sequestration (even if this may not be the most important way to increase carbon dioxide absorption globally). However, the extent of carbon dioxide absorption depends on the net biomass production in the land. Though there are other benefits of restoration, most of these can be achieved when land is restored with the previously mentioned three primary objectives.

Need to Focus on Marginal Benefits and Costs

When we discuss the viability of an action towards restoration, there is a need to focus on costs and benefits of the action-more specifically, the marginal costs and benefits. Certain actions which are costlier to society may become beneficial when this 'marginality' is taken into account. Let us take an example to make this point clear. If a part of a natural forest is cut down, and then a plantation of indigenous trees is created, there is a definite negative impact in terms of land restoration. On the other hand, the creation of such a plantation in an already degraded land will have a positive benefit. Hence, it is not correct to focus on the absolute costs or benefits of different means of land restoration.

Trade-Offs Between the Protection of Biodiversity and Addressing Climate Change Through Land Restoration

If the objective of land restoration is to preserve biodiversity, then it is to be protected as a mature or secondary forest (or other ecosystems, such as natural grasslands, wetlands, etc.). By protecting it, a pool of species and genetic diversity is conserved. If such a pool is already conserved in say 50,000 hectares, then the additional gain in having the same pool in another 10,000 hectares is not that high. There can be a declining marginal value in protecting the same pool of diversity as the area of land increases (probably after a threshold limit). This feature implies that specific pools of biodiversity need to be protected, but it may not require the protection of the same forest type beyond a specific area. The remaining part of forests or land which can sustain the same pool of diversity may be managed for other purposes.

One such purpose can be the absorption of carbon dioxide, even though it is not necessary to view the protection of natural forests and tree plantations as the only or major tool to reduce carbon dioxide emission. If the absorption of carbon dioxide is the objective of restoring land, there has to be an appropriate management for this purpose. This is to be in such a way that the additional biomass production in each year should be significant so that the absorption of carbon dioxide is also significant.

A mature forest may not have too much additional biomass production and hence, may not absorb carbon dioxide notably. Tropical forests may achieve this maturity (peak in terms of biomass stocks within 100–120 years). Hence, the conservation of secondary forests in a manner that does not promote the attainment of maturity is important for sustaining the benefits of carbon dioxide absorption.

Nevertheless, the management of secondary forests, which implies pruning or harvesting parts of biomass (like timber) regularly and the use of these in a manner that avoids the release of carbon dioxide back into the atmosphere (for example, cutting down a part of the trees for timber, and using it for making furniture or constructing houses are desirable with this goal), can enhance carbon sequestration in forested lands. This process will ensure that the annual absorption of carbon dioxide remains higher even after the period that is required for the forest to attain maturity. Other goals of protecting forests, like the maintenance of a land cover, preservation of soil and water, etc., can be achieved with such a managed forest.

If we take the conservation of biodiversity and absorption of carbon dioxide as two goals of protecting forests, the pursuit of one may affect the other, and hence, it is better to pursue these in different areas. This is similar to the case for zoning where it is efficient to carry out two activities in two separate spaces rather than in one.[3] One part of the forest is to be preserved (untouched) for biodiversity, and the other can be managed to enhance the absorption of carbon dioxide. This means that there should not be a greater expectation of protection of biodiversity from those forests which are protected for carbon dioxide absorption and vice versa. Though this is a common strategy in the management of forests, many environmentalists argue for the extension of fully conserved areas without understanding these costs and benefits.

Plantations of one or more trees can also help in absorbing carbon dioxide if the harvested material is not burnt. Whether to have one or more crops can be based on concerns about efficiency and the value of biomass. There is no point in bringing in the concerns about biodiversity in such plantations (since these are not appropriate for protecting such diversity anyway). One can evaluate plantations based on the vegetative cover over land and their ability to conserve soil and water, but it is somewhat pointless to evaluate them based on the goal of biodiversity. If we accept this position, there is no major difference between plantations of one species and multiple species if these are adequate to conserve soil and water resources (and if the harvested material is not burnt and carbon dioxide is not released back into the atmosphere).

[3] This is discussed in economics as the non-convexity of production possibility frontier. For example, fishing and tourism may use the seaside but it may not be efficient if both these activities are carried on at the same beach. Instead, one part of the seaside can be reserved for fishing and another for tourism.

The Real Issue with Exotic Species

If plantations serve the purpose of absorbing carbon dioxide and providing a sustainable vegetative cover over the land (to conserve soil and water), then one needs to consider the appropriateness of exotic species for this purpose. In certain cases, exotic species like Eucalyptus are already in the country, and preventing them on the basis of indigenousness, per se, is not useful. However, whether the propagation of such exotic species in plantations can 'over-compete' the indigenous species in nearby territories could be a major concern. This can be a consideration in avoiding plantations of certain exotic species. If there is no such problem, there is merit in the plantation of exotic species as one means of land restoration (when these can sustain vegetative cover and conserve soil and water resources).

It is possible that plantations of certain exotic species may have a negative impact on soil and water conservation. The traditional concern about trees like eucalyptus is their absorption of water. Whether this concern is still relevant in the new generation eucalyptus plantations which use seedlings with a shallow root structure needs to be taken into account.[4] If not, the advantages of cultivating a commercially valuable tree, like eucalyptus, and harvesting it without destroying the land cover could be an attractive strategy, especially, in countries like Brazil where land is not a scarce resource.

An appropriate strategy for land restoration also depends on the incentive for individuals and firms. Let us think about the regeneration of natural (secondary) forests as one restoration strategy. If this strategy does not provide any significant direct benefits to individuals (and hence, society), then people/society may not follow such a strategy. This strategy, then, becomes impractical or unviable. Different kinds of private incentives have to be considered in deciding the trade-offs between strategies or means of land restoration, discussed next.

Private Incentives for Land Restoration and Their Impact on the Selection of Means

Private individuals who invest in land restoration may have different interests or motivations. There may be individuals who invest in land restoration without expecting any monetary gain. (Though these interests determine public interest too, we may consider these public or altruistic choices or not so rational reasons in another chapter.) The joy of seeing that the land is restored itself may be adequate to motivate them. It may be noted that these individuals who are investing in restoration incur a cost for this purpose but there are no financial revenues. Hence, they may encounter a net financial loss which may be compensated by the non-monetary gains from restoration. They may be getting finance from other investments which are

[4] https: //www.downtoearth.org.in/blog/water/why-eucalyptus–60275.

not related to land. There may be wealthy individuals who can devote a significant amount of resources to this purpose. (There may be other people who can devote limited resources for this purpose). For these individuals who are driven solely by the environmental benefits of land restoration, natural regeneration (or the establishment of secondary forests) can be attractive.

For some of these individuals, these non-monetary gains may include their contribution to the reduction of carbon dioxide in the atmosphere. (Some of them may use this carbon sequestration as a way to gain revenues from carbon finance. However, others may be interested in social gains from carbon sequestration.) If this contribution has to be sustainable, then biomass production is also to be sustainable. This may require pruning and harvesting of vegetation from the restored land. Some of the trade-offs mentioned in the previous section are relevant in this regard. The main point is that there could be trade-offs even for those who restore land mainly (or only) for environmental/social benefits. However, such gains (or motivation to derive such benefits) on the part of a set of individuals may not be adequate to restore or conserve the whole area of land which is prone to degradation.

There are many other individuals and private organisations which are interested in a monetary return from land restoration.[5] This expected return may vary between that which is equal to the market return from any financial investment to zero per cent. There may be other individuals who want to recover a part of the cost of restoration.

Monetary returns from restored land are important for both these sets of people. Even for these people, restoring natural vegetation or establishing secondary forests can be attractive as long as revenues from sources, such as non-timber forest products or eco-tourism, which can be obtained without disturbing the nature of vegetation, are notable. However, the net income from these sources depends on the socioeconomic and geographical contexts. In certain places, wildlife viewing is possible without harming natural plant growth and it can be a major source of revenue.[6] However, the availability of more land of this kind may reduce the returns per land from this economic activity. Non-timber forest products (NTFP) can be an important source of revenue but here we need to reckon the cost of collection (Zashimuddin, 2004). Hence, it may generate higher incomes where the cost of collection is lesser (probably where the cost of labour is less as in territories which poorer people inhabit). There may be contexts, such as Brazil, where the income from NTFP or eco-tourism may not be significant. Given the availability of a lot more land, these incomes even if significant, cannot be a major incentive to restore a major part of the land in such countries.

Hence, monetary gains from harvesting timber and non-timber products in lands undergoing forest restoration may serve as a private incentive for land restoration. The discussions in the previous section indicate that this can be achieved if the goal

[5] There are initiatives of this kind in not only the USA but also in developing countries, such as Brazil.

[6] This is the case in parts of Africa. https://sustain.pata.org/wp-content/uploads/2014/12/WildlifeTourism-impacts.pdf; http://www2.dnr.cornell.edu/saw44/NTRES431/Products/Spring%202003/Module2/Ecotourism.pdf.

of biodiversity is not pursued as part of land restoration (or if that part is taken care of by the government in national parks or legally mandated areas in private lands so that species/gene pools of all types can be conserved). However, other goals of land restoration, such as the absorption of carbon dioxide and the prevention of soil erosion can be achieved by sustainably harvesting part of the biomass (like timber) and using it in a manner that does not release carbon dioxide back into the atmosphere in the short term.[7] Moreover, part of the land restored by private individuals or firms can be used for restoring natural (secondary) forests. This can be done as part of mandatory requirements (as in the case of Brazil) or voluntarily for the non-monetary gains mentioned earlier.

The other case is to explore the possibilities of land restoration in commercial plantations and agricultural farms. There have to be changes in plantation/cultivation practices (and the development of technologies for this purpose) to see that these do not lead to the degradation of land. This may include changes to ensure that the land has vegetative cover throughout the year, excessive chemicals are not used, water/moisture is conserved, organic nutrients/minerals are not destroyed, and so on.[8] It is also important to examine if these changes increase the costs and/or reduce the productivity[9] and consequently, reduce the profits of plantation/cultivation. If this additional cost is insignificant, then land restoration in such areas may not be a major challenge. If it is significant, this is the additional cost to be borne by private landholders for the non-monetary gains from land restoration. A sizeable section of landholders may be able to bear this cost if adequately motivated. A subsidy that can compensate for this additional cost should be able to motivate most of the landholders to move towards land restoration. (Technological changes can reduce the amount of subsidy that is required for this purpose too).

Conservation Strategies in Public Lands: A Normative View

The conservation of public lands should reflect the priorities of the population, as a whole. If we include future generations also, this conservation has to be driven by sustainable development goals.

Since the ownership of these lands is with a single entity (the government) representing the whole society, the problem of collective action can be solved relatively easily (This becomes clear when we compare the effort that is needed to persuade millions of individuals and firms to adopt a restoration strategy). Hence, it is cost-effective to use public lands for those restoration goals for which private benefits are much less when compared to social benefits. Based on this framework, it is better if public lands are used for goals such as the conservation of biodiversity. This is

[7] This is the strategy of firms like ReGreen or Symbiosis in Brazil.

[8] Innovations in organic agriculture aim at these changes. Https: //www.fao.org/organicag/oa-faq/oa-faq6/en/.

[9] This is a concern about organic farming. Refer Gilbert (2012).

so since the conservation of land with this goal may not give significant income to individuals (and the present generation) even though these are socially useful.

When public lands are conserved with the goal of conserving biodiversity, other lands can be conserved with goals that ensure the restoration (or the avoidance of degradation) with certain financial returns. Hence, the use of such land for the sustainable harvest of timber or for the optimal absorption of carbon dioxide (which may require the management of biomass production) becomes part of a socially useful strategy.

Conclusion

The restoration of land should happen everywhere. However, there are other purposes for which land has to be used. Hence, land restoration cannot happen with any one specific goal, or with a goal that makes the utilisation of land difficult for other purposes. It is important to pursue land restoration with different goals in multiple parcels of land. There can be trade-offs (as noted earlier) when these different goals are pursued and land restoration in a country as a whole needs to take these trade-offs into account and devise different strategies in different locations so that the overall goal of restoration is achieved. This is much more important in populous countries in south and east Asia or parts of Africa or other developing countries where agriculture or related activities continue to be an important occupation for a significant share of the population.

The discussion in the chapter indicates the importance of integrating the insights of ecological and natural sciences and economics into the design of land restoration projects. This is important since there are not only benefits of land restoration but also costs (including opportunity costs). These costs may vary for each strategy of restoration which in turn means that there are trade-offs between different strategies of restoration. Hence what is optimal for each individual or group or society or a country may be different and there cannot be a universal best strategy. Science can change these possibilities (opportunity set). For example, if an exotic species can establish land cover quickly but it may lead to an over-extraction of water, then genetic modifications that lead to a change in its intake of water can help and affect the trade-offs which we have mentioned earlier. By bring these outcomes of science and values (of benefits and costs) which are identified by economics together, an optimal strategy of land restoration for each piece of land is to be selected.

Chapter 8
Financing Land Restoration: Possibilities and Challenges

There are different sources of finance for land restoration. However, the extent of the money available and their possible impacts may vary. We consider these different sources in detail in this chapter. Rather than taking an empirical approach of the money that is available for land restoration, this chapter takes a theoretical view of the potential and limitations of each source of finance. This discussion is also connected to empirical situation from different countries.

There are overviews on financing land restoration (for example, Sophus et al. 2024 and Ding et al. 2017). However this chapter takes a step-by-step approach to discuss the possibilities and limitations of different sources of finance.

One way to classify these sources of finance are private versus public. Within private, there can be individuals, companies or altruistic organisations. Within public, there are national and international public organisations. Another way to think about the source of finance is to connect them with different motivations. There can be for-profit actions of individuals and private companies which can lead to land restoration. However there can be altruistic goals including ecosystem conservation which may drive the actions of even individuals and private companies, and these can help land restoration. There can also be mixed goals with profit-enhancement and environmental conservation, driving actions of different entities.

Private Sources

There are individuals, companies and altruistic organisations under this category. Each can be motivated by profit, altruism or mixed goals. Let us take these different types.

V. Santhakumar et al., *Land Restoration: A Rational, Grounded and Pragmatic Approach*, https://doi.org/10.1007/978-3-032-26652-1_8

For-Profit Actions of Individuals

Certain kinds of land restoration is possible even when profit is the driving force of individual action. For example, farmers may take steps to control soil erosion (since they may get a major part of its benefits and this part may be greater than the cost of these steps). Though these not are adequate to land restoration at socially efficient level. This is due to the positive externality. Their actions of land restoration may neglect the benefits enjoyed by others due to this positive externality. Hence the same farmers who take measures to control soil erosion may not take adequate steps to enhance carbon sequestration through plants/trees in their land. However, even rational profit-seeking individuals may attempt certain kind of restoration. For example, if the degradation of land is leading to a decline of its fertility with a negative impact on agricultural yield (and this is indeed an outcome of land degradation), then informed farmers have an incentive to arrest such a degradation.

Altruist Goals of Individuals

In our interactions with number of individuals who are involved in land restoration-these include NGO leaders, private company managers interested in land restoration, and also those who buy land and allow it restore - we could see a set of them driven by ecosystem restoration, or by combining such restoration to their profit motives. For example, a set of individuals bought about 100 acres of land near a nature reserve and allowed it be forests.[1] There are individuals who have left their own land (which has undergone different levels of degradation) uncultivated for natural regeneration or used a variety of methods to bring back indigenous trees and plants. They may be using a variety of activities to use this land (including ecotourism) to generate income or to sustain public interest in their actions.[2] There are wealthy individuals who buy large areas of land (for example in Brazil) and allow these to develop into natural forests. The wealth of a set of rich individuals are used to buy land and 'return' to nature. Using private lands for conservation purposes is an accepted practice globally.[3] However all these constitute a small part of resource allocation for land restoration.

[1] https://navadarshanam.org/.

[2] Another such case is described here: https://greenara.in/.

[3] For example, https://landtrustalliance.org/.

For-Profit Private Companies

Usually those companies which use land for their different business opportunities may not invest in restoration if there are no government mandates for this purpose. There are private companies which plant trees for say, paper production but their goals may not lead to sustainable land restoration. However with regulations enforced by governments and relatively minor changes in practices, these companies can also facilitate land restoration to some extent. The goals of these companies may change based on the goals of its shareholders or even consumers. However, over time, we can see private companies which see land restoration as a business (and we discuss this in detail in another chapter). These may not be fully profit oriented but have certain conservation goals also. It attracts certain level of capital for restoration. Those who conserve land by expecting to gain from carbon finance is also under this category. This is an important part of private restoration finance.

Social Actions of Private Firms

Private companies are also involved in social and environmental actions. These can be partly aimed at reputation building. Or it can be a response towards government regulations. For example, Corporate Social Responsibility (CSR) in India is of that category.[4] We could see companies using CSR funds for land restoration activities. There can also be intrinsic motivation on the part of corporate owners or managers encouraging them to invest in ecosystem restoration and other social purposes. Though this can be a source of finance, its size is meagre compared to the needs of land restoration globally.[5]

Altruistic Organisations

There are philanthropic organisations which fund activities including ecosystem restoration.[6] The origins of their endowments are (and these may be invested in) for-profit activities, but the objectives of philanthropies is not the maximisation of profits. There are certain advantages of these organisations. Usually they invest from the returns of endowments and hence investible resources are available for a very long-time. The can have very long-term goals also, and in this way, are different from

[4] The Companies Act of India, 2013 recognizes environmental conservation as a CSR activity and eligible profitable firms have to invest a certain share of their PAT on CSR activities.

[5] For instance, in India, CSR spending is often concentrated on specific projects and may not address the larger systemic issues of restoration. Source: https://icar.org.in/sites/default/files/2022-06/ICAR-CSR-FUND-GUIDELINES.pdf.

[6] For instance, the Nature Conservatory, The Conservation Land among others.

governments whose goals may change based on the parties in power. Philanthropies can invest their resources for innovative models and pilot experiments, and lessons from these can be used to scale up restoration projects. Of course, there are a variety of social goals which are pursued by these philanthropies, and environmental conservation is only among them. Even within environmental conservation, land restoration may not be at the top in agenda.

Public Sources

There can be national governments and international organisations under this category. They may directly provide loans and grants. They can also extend subsidies to change the actions of individuals. It can also create 'markets' and the payment of ecosystem services is one such case. These are described in the following sections.

Governments

Grants and Loans

Individuals may not take adequate measures towards restoring their land due to a variety of reasons which we have described earlier. Hence it is important for provincial and national governments to take a higher burden in this regard. Conventionally, governments in different countries have used grants and loans to support agriculture. However there is a need to reorient these to facilitate land restoration. This is already happening in different countries. Some governments may take international grants or loans but distribute these to land owners to move away from land degrading practices. We could see this happening for the peatland conservation in Indonesia. Brazil is also using different tools (including subsidised interest rates) to encourage land restoration.

Subsidies

Governments provide a variety of subsidies for different social purposes, and some of these have implications for land. We have noted the possible negative impact of subsidies on fertilisers, pesticides and irrigation water on land in previous chapters. However subsidies can be used for land restoration also. There are real world examples too. China adopts a policy to pay farmers who do not cultivate (certain crops in certain contexts). India has come out with different policies to facilitate organic agriculture. There are subsidies for soil and water conservation (which may help land restoration) in many countries. These may depend on the financial resources which are available with governments and the democratic compulsion to follow these policies vis-a-vis other expenditure programs. In the long-run, national governments have to be the most important source of finance for land restoration within each country, and this may manifest in more subsidies favouring land restoration.

Payments for Ecosystem Services

Land restoration has positive externalities which benefits not only those invest in restoration. Most of these externalities such as reduced social erosion or increased water conservation benefit people within the nation (though there are global positive externalities which we discuss in the next section). Hence it is important for national governments to ensure that a part of the benefits that society gets, is transferred to those who invest in land restoration, so that they have adequate incentive to consider these wider social benefits also in their choices. That is the purpose of payment for ecosystem services.[7] For example, when farmers in the upstream of a river or reservoir take soil and water conservation measures, it can benefit the downstream users, but there has to be a mechanism to see that upstream farmers gain from these downstream benefits. There can be a transfer mechanism for this purpose. If there is a hydroelectric power station which uses water from a reservoir, it can transfer money to the land owners in the upper catchment of the reservoir so that they have an interest in practising soil and water conservation. Such payment systems are used for ecosystem conservation in Costa Rica and also for water conservation in the US. It is advocated as a source of finance for land restoration, though it is yet to be adopted on a wider scale in many other (less developed) countries.

International Organisations

Grants and Loans

The increased awareness on the importance of land restoration for its own sake (say, to combat desertification) or for a variety of purposes including the mitigation of climate change and conservation of biodiversity, has encouraged developed countries and international organisations (over which these countries have greater control) to allocate money to provide grants and loans to less-developed countries for this purpose. This may be for the reduction of the negative externality associated with land degradation. In that case, it becomes a global public good for which international organisations provide financial assistance. There can also be international altruistic organisations like global philanthropic foundations which may also provide financial support for the same purpose. It is not only the traditional aid-givers (like Scandinavian countries) but also newer ones like Saudi Arabia are at the forefront of supporting other countries for land restoration. This is an important source of finance for global land restoration.

---

[7] For a review of the literature on this instrument, see Tuyet-Anh et al. (2024).

Payments for Biodiversity Conservation

Biodiversity conservation provides global benefits. Since all kinds of biodiversity cannot be preserved in one or few developed countries, there is an interest in its global conservation. This is also creating an interest in developed countries to financially support the preservation of biodiversity in other countries. This is part of fulfilling their responsibility to protect biodiversity by funding other countries which are yet to destroy such diversity. There are different mechanisms for this purpose (like those for addressing climate change). This can be an important source of money for land restoration, since a part of (not all) projects for this purpose may enhance biodiversity.[8]

Carbon Finance[9]

The money that is earmarked to address climate change can be used for land restoration. The potential for carbon credit is higher in this regard, especially for less-developed countries. This is so since most developed countries are highly industrialised, and the marginal cost of reducing carbon dioxide emissions in these countries could be higher. However they can buy carbon credits from poorer countries where the marginal cost of reducing CO_2 emissions or absorbing CO_2 could be lower. Land restoration through different kinds of vegetative growth can be a way of having additional absorption of CO_2, and this can be a relatively cheaper investment in less-developed countries where land is available. This will benefit these countries through the restoration of land, which could be an important source of livelihood. Theoretically carbon credits can be a notable source of finance for land restoration, though there are practical challenges which we discuss in the following section.

Challenges

We discussed different sources of finance that are available for land restoration in the previous paragraphs. However these may have certain generic challenges which may work against the availability of adequate money that is necessary for land restoration in the world as a whole. Some of these limitations are discussed in the following sections.

---

[8] There is a distinction between direct and indirect finances for biodiversity protection. Refer Ferraro and Kiss, 2002.

[9] For an accessible review of carbon finance, refer https://www.renewableinstitute.org/carbon-finance-explained-everything-you-should-know/; the potential of carbon market for land restoration is discussed in the report https://www.endangeredlandscapes.org/wp-content/uploads/2023/07/ELP_Carbon-Finance_Practitioners-Primer-v2_JULY_approved2.pdf.

Majority of Individuals May Not be Interested in Investing in Land Restoration

It is a reality that most individuals are not interested in investing in land restoration. This reluctance is privately rational when we note that land restoration has positive externalities (which means that a notable part of benefits of such investments may go to society as a whole and not only to the investor). The conservation of biodiversity, absorption of carbon dioxide, increased ground-water storages, and so on, due to the conservation of a stretch of land (controlled by an individual) are such positive externalities. When these other beneficiaries cannot be charged a price, the investor in land restoration will not take into their benefits, and will be underinvesting in such restoration. Though this is the general picture, there can be other reasons also. Many farmers (mostly in less-developed countries) can be irrational and hence may discount heavily the future gains. These can be other behavioural traits which we discuss in another chapter. There may not be enough altruistic investments for social and environmental purposes from individuals.

Governments in Poorer Countries May Not Have Enough Resources to Invest for Land Restoration

When individuals are reluctant to invest in land restoration due to positive externalities, it should be the role of government to invest for such purposes. National governments are able to capture positive externalities within a country. It is true that land restoration has global externalities through the absorption of CO2 and conservation of biodiversity, and each country may not benefit fully from these externalities. Hence there could be a reluctance on the part of each country to invest for global positive externalities. However, we may see different countries unable to invest adequately even to capture national positive externalities. This is more so for less-developed countries. There are multiple public investments which have such national positive externalities. Education and healthcare are two examples but many of these countries are unable to invest adequately for these purposes also. The tax resources available to these countries may not be enough for these investments. There can be a political economy issue, since a part of the return from investments in land restoration and other ecosystem services may go to future generations, and they are not present currently to influence the policies. Hence governments are not compelled democratically to invest adequately for restoration purposes. It can also weaken the enforcement of environmental regulation (an issue that we discuss in another section).

Majority of Private Companies Are Not Interested in Land Restoration

Though there is a small set of private companies which invest in land restoration as part of their social responsibility, and another set which see such restoration as a business, the majority of private organisations is neither investing not interested in land restoration. This is true even in the case of those companies which use land for one or other purposes. Even when there are regulations on how to use land (restore it after use in the case of mining and such activities), these are not adequately enforced in many countries. The lack of adequate resources on the part of governments, an issue mentioned previously, could be the major constraint. In addition, the lack of awareness, corruption, not enough pressure on the part of citizens, etc. may weaken the enforcement of regulations. When private companies want to show commitment to environment for reputation, land restoration is not an 'attractive' investment.

Transaction Costs Higher for a Large Number of Small Projects and Such Projects Are Common for Land Restoration

Global finance mechanisms may allocate money in millions of dollars for land restoration. However land restoration in most cases, especially in poorer and less-developed economies with higher population densities may happen in tens or hundreds of hectares at the most. These require smaller amounts of money. These projects may not be attractive to funding mechanisms which have millions of dollars. This is primarily due to the transaction cost or processing costs of small projects. Even a small project may require higher cost for due diligence and contracting, and also for monitoring and evaluation. Hence these funding mechanisms may prefer large scale projects. Bringing together land-holders for a large scale restoration project may encounter coordination problems. (In developed countries, large stretches of land may be owned by one or other individual but that may not be the case in most poorer countries.) All these may work against the formulation of large scale projects which can attract funding from international organisations.

Many Actors Do Not Have Enough Capacity to Access Carbon and Biodiversity Finance

Theoretically one can say that markets for carbon dioxide absorption and biodiversity conservation can be used to finance land restoration projects. However there are only very few cases where these projects actually get these money. There could be two issues: First is that the procedures required for recording the additional contribution towards CO_2 absorption and claiming carbon credit could be tiresome and costly and this may discourage many small-scale projects of land restoration. Even for large-scale projects, the human resources which are required for this purpose (of documenting and claiming carbon credit) is substantial. Many less-developed countries do not have such trained people at appropriate positions. Since large scale projects in land restoration (say through forest conservation) may happen through government initiatives, the absence of an adequate capacity among government officials can be a major constraint. We have noted such a problem while looking at the peatland restoration in Indonesia.

Mixed Goals Private Finance—The Size Is Still Small

We have noted that there is an emergence private finance for land restoration, and much of this is not solely profit-oriented. There is an interest in investing for sustainable development goals. However, the size of this investment is too small compared to the overall requirements of land restoration. In countries where such businesses operate, they struggle to get more capital to scale up their business. We have noted such a case in Indonesia. The amount of private capital that is available for this purpose is severely inadequate for the land restoration needs of such countries. Moreover, since these investors aim at receiving certain revenue, there are many needs of land restoration which may not provide such revenues, and they may not be able spend money for such restoration needs. There is one more limitation in the reliance on this private capital and we discuss it in the next section.

Private Finance May Go to Locations Where Institutions Are Reasonably Functional

We have noted the emergence of private finance which see land restoration as a business and/or environmentally sound investment. However, these may not reach adequately to locations where there is a higher level of land degradation (possibly due to development, institutional and governance issues). This is similar to the flow of private finance (including foreign capital) for normal economic activities. It is noted that such a flow is shaped by the 'ease of doing business' in a country. However the 'ease of doing business' for land restoration business could be a lot more problematic. For example, ensuring property rights over a longer period of time (or policy stability over a longer period) is necessary for private firms which create forests or indigenous plantations in stretches of land to recoup their investments. Many less-developed countries (which are democratic or non-democratic) may not be able to communicate such signals of policy stability. Moreover the regulations against ecosystem degradation should be stronger. Otherwise, those who invest in sustainable production may have to compete with those who degrade ecosystems in certain markets (say, of timber or such products). This too may reduce the attractiveness of certain countries to private investors who see land restoration as a business.

Effectiveness of International Altruistic Investments May Be Lower in Certain Contexts (Mainly Poorer Countries)

Land restoration in many countries, especially less-developed countries, is funded by international altruistic investments. These could be public or private (say philanthropic) organisations. For example, there was a support from Norway for the peatland restoration in Indonesia. One can see such a support for the Great Green Wall initiative in Africa. However there are challenges in the realisation of the full potential of this finance. There can be local factors (which cannot be solved merely through the finance) which may work against land restoration in these contexts, and this may discourage these funding sources. We have seen such a situation in the context of Norwegian funding to Indonesia. These local factors could be connected to the development, institutions and governance of these countries. Or international funding agencies may neglect the impact of these factors on land degradation or restoration which can make their funding programs less-effective. Hence even when there are declarations of financial allocations at the international level for land restoration in poorer or less-developed countries, this money may not reach such countries adequately.

Chapter 9
Need to Reckon the Cost of Land Restoration

The economic benefits of land restoration are widely known in policy and academic circles. Conversely, financial costs of land degradation are also known among policy makers, researchers, etc. However the social costs of land restoration are not adequately understood.[1] One can argue that these costs in aggregate are lesser than economic benefits of restoration. That is true in most cases. However these who gain these economic benefits may be different from those who may have to bear the cost of restoration. There can be those who are net-gainers and others who are net losers (even after considering their benefits). The latter needs special consideration of society. Moreover, the presence of a section of society who are net losers can lead to a distributional and political economy struggle around restoration and it can delay the projects/programs for this purpose. Hence an understanding of social costs of land restoration and who have to bear these, is important to design distributional or compensatory policies which are necessary to accelerate the process of land restoration. The creation of such an understanding is the objective of this chapter.

There are aggregate cost estimates of land restoration such as in Verhoeven et al. (2024). Rather than giving such estimates, this chapter discusses the different types of costs and their importance.

Direct Costs

The cost of land restoration may include direct and indirect ones. For example, the cost of establishing vegetative cover over land is a direct cost. In certain cases, restoration may require the prevention of all human activities in a stretch of land, so that it may be allowed to have the natural growth of vegetation. However, there are substantial direct costs for this purpose, even though vegetative cover comes

[1] It is noted that the costs of restoration of ecosystems are less frequently analysed. For example, Ma et al. (2020).

V. Santhakumar et al., *Land Restoration: A Rational, Grounded and Pragmatic Approach*, https://doi.org/10.1007/978-3-032-26652-1_9

without bearing any direct cost. These may include administrative costs for strict policies that prevent human activities (in that stretch of land), and also for its effective enforcement. There may be cases where such natural regeneration may take a much longer time, and there may be a need for reintroduction of indigenous plants and trees. There can be a cost for these interventions. However these direct costs of land restoration are understood reasonably well and usually taken into account in project estimates or government policies.

Indirect Costs

There are indirect costs of land restoration. These are mostly connected to opportunity costs—the economic benefits foregone due to the restoration of a stretch of land. A greater appreciation of these indirect costs are needed to predict the effectiveness of restoration projects.

The most important cost is that to be incurred for not using land in a manner that may increase degradation. If there is partial or full restoration, then land should not be used at all, and that can be very costly for a significant share of the world population. There can be less degrading ways of using land but that may also require forgoing certain benefits. What is more important is that these costs may vary for different people and different contexts. Let us elaborate these with a few examples.

If the land is used for short period crops like food-grains or sugar cane or oil seeds, the level of degradation could be higher. This is due to the removal of land cover once in a few months or a year, and the tilling of soil before every cropping season. However the sacrifice of this crop production for land restoration can be costly for specific contexts and people. Ideally, only a small part of the land is needed to produce all these outputs for the global population if their productivities are higher. However there are a couple of challenges in this regard. Poorer people who may not generate much cash income may use their labour to generate these outputs in the land that is accessible to them (since they cannot buy these products from markets). This has a higher possibility if the governments are not in a position to subsidise their consumption of these products. This is partly because there are not many opportunities where they can use their labour and earn an income to buy these products from markets (where these can reach from high productivity areas). Hence even if the productivity of the production of crops is higher in some other contexts, it may not discourage people from other (poorer) contexts from their cultivation. Even if the productivity of these outputs in a few countries are higher, that may not reduce the need to produce these in other countries where productivity is lower, due to different barriers in international trade for farm products.

We can consider another indirect cost of land restoration. The use of biofuel for cooking and heating needs can also be a source of land degradation. This was found to be a major cause of the disappearance of trees in parts of Central Asia after the collapse of the Soviet Union. People may not have the resources to buy and use less-harmful cooking and heating fuels. The transition to these alternate fuels is costly and this may have to be borne by the government. Or land restoration in these areas may require the (subsidised) supply of other fuels and this can be costly.

The non-use of land (and the diversion of land away from cultivation or extraction of minerals) may have implications at the national or global level. This may be reflected in lower production of products (say farm products) for consumption and also inputs (like minerals) for industrial processes. Let us take one product which contributes to consumption directly and the production of other outputs: energy. Most of the conventional ways of producing energy—that which requires coal or petroleum or hydroelectric plants—need the use of land. All these ways of using land can lead to its degradation. Moreover, even the non-conventional or renewable ways of producing electricity like wind turbines or solar power also need the use of land. It is noted that the expansion of solar energy (which is considered environment friendly) also may need extensive use of land, and it can also lead to the degradation of terrestrial ecosystems. Moreover certain rare minerals which are used for the mass production of solar panels may have to be extracted, and some of these pose greater environmental problems in the surrounding areas. Hence any reduction in the use of land for production of energy may increase its price and that is a cost that the society has to bear for land restoration. This may be the case of a number of other products. Moreover, a significant share of people in poorer countries are employed or find their livelihood in agriculture. The transfer of these people to industrial or service sectors has not been that smooth in these countries. There could be two issues: First, many of these people may not have the human capital to get jobs in industrial and service sectors. Secondly these countries may not have capital investments to create enough jobs to attract those people who need to get out of agriculture and other land-based occupations.

For the supply of agricultural products, the increase in productivity may reduce the need for land for this purpose to some extent. Usually developed countries use only a small part of their land (and labour) for their farming (and a major part of the remaining land is restored). However these countries also import a significant share of these products from less-developed countries where the productivity may not be high and are using a notable share of land for this purpose.[2] Moreover the increase in consumption (partly due to population growth but mainly due to income growth) in different countries enhances the requirement of land for this purposes. It may be noted that the increase in the consumption of meat-protein (which is an indicator of increasing incomes of people) may also require the diversion of a greater share of land for the production of animal feed. Hence the land that can be diverted away from agriculture (for natural vegetative growth) is also limited. Or such a diversion may increase the cost of consumption in different countries. There may be a need for investment to see that a lesser share of land could produce all farm products which are necessary for human (and domesticated animal) life.

One may argue that there are opportunities to restore land without reducing its use for agriculture. Organic farming is touted as one such method. However we need

---

[2] *Source* https://azimpremjiuniversity.edu.in/insights-from-social-context/socioeconomic-conditions-that-enable-restoration-of-land-and-ecosystems-what-we-know-what-we-need-to-know.

to be concerned about two-three costs here: First, available evidence indicate that the yield rate of organic farms is lesser than those which use chemical inputs.[3] This would mean the use of a higher stretch of land if the same amount of farm product is to be supplied through organic agriculture. A related point is that the price of organic products is higher partly reflecting its lower yields. If a major share of agriculture is becoming organic, there could be a consequent increase in the price of farm outputs and hence the cost of consumption. We cannot be sure whether all inputs that are used for organic agriculture will not lead to land degradation or not. It uses land, plant-based inputs (like organic manure) or the waste from animals (which also use plant materials) and an increase in the use of these inputs can also accelerate degradation.

Even without organic agriculture, there can be careful practices of agriculture which may reduce degradation of land. However these may also increase the cost of agricultural products which may reflect in their prices. There may be a situation where consumers may have to pay higher price for land restoring agriculture and associated practices. This price is to be recognised as a cost of land restoration.

Similar to organic farming, productivity can be lower even in non-degrading agriculture making it costly to practice. We have seen such a case in the wetlands of Indonesia. Burning is the cheapest method for farmers or companies to make peatlands conducive for cultivation (or plantations like that of oil palms), even though such burning has negative externalities. Hence any change in the method of preparing peatland soil (with the purpose of reducing land degradation) can be costly to those who use it for cultivation. They will not change the practice unless their higher cost is compensated or tight regulations compel them to do so. The cost of enforcing these regulations can be higher when private incentives are geared towards burning peatlands. When a sizeable section of people are of this type, political economy may also work against the proper enforcement of regulations, and we discuss this issue in the next chapter.

It is possible to have a situation where most people voluntarily quit small-scale agriculture and land-based occupations. This has happened in the developed countries. However this requires, as noted earlier, people to acquire education and skills to depend on industrial and service sectors for jobs. Hence another cost of land restoration is that for human development (including educational development). However human development has many other benefits, and bearing this cost is good for societies, and it is good if it leads to land restoration also to some extent. However for countries which are currently at a lower level of economic/human development, the cost of achieving higher levels of such a development for voluntary non-use of land, also needs to be accounted.

We have noted the possibly higher price of agricultural products when land restoration practices become common. This kind of a development is not uncommon. For example, in order to address climate change, certain practices like the use of coal may have to be reduced and this may lead to an increase in the price of goods (like energy). However, technology may help to reduce the cost of alternatives like the renewable energy. Though this should be possible in the case of land restoration also,

---

[3] *Source* https://mankindag.com/blog/difference-between-organic-farming-and-chemical-farming.

the challenges are much higher. This can be due to different reasons: (a) the large share of land which has undergone degradation and the need to restore it; (b) the large number of small-scale producers who make these products and for whom there could be severe challenges in the adoption of technology; (c) technological change in agriculture has taken place to a great extent, and so far it has not been that enabling land restoration. In order to restore a significant share of land without affecting human consumption, there may have to be radically different kinds of technological changes. (For example, what if people get enough food from natural growth of vegetation in different parts of land?) However such technological changes cannot be foreseen in the current context.

Costs for Specific Sections of Society

What we have described so far are aggregate costs. However there can communities which depend on land and whose activities may have to be regulated for land restoration may be poorer and less-privileged. This may be the case in many less-developed countries. In general, people who are involved in agriculture are economically vulnerable to do the lower income elasticity of farm goods. (The relative price of farm products may not go up like that of industrial and service products when the overall income in society grows).

There are also landless workers (with almost no endowments other than their ability to do physical labour) in agriculture. Certain restoration strategies can affect them the most, and the governments may not compensate them adequately. We have described the situation of tribal communities in India elsewhere. They were depending on forests conventionally but their rights were controlled by successive governments. Though this has enabled the protection of forests to a certain extent, it is achieved at a higher social cost and it remains uncompensated. In many parts of the world, indigenous people live in areas where biodiversity is higher. Amazon basin is an interesting example but there are other equally important cases. Borneo in Indonesia is another case. To some extent, this higher level of biodiversity in their areas is an outcome of their less interventionist use of land over time. However they may have to pay a price for this currently if their access to or use of land is restricted or moderated for the global benefits of land restoration. This is not an argument against land and ecosystem restoration in these areas, but to highlight the need to compensate these people to have a decent life even with such restrictions of their use of land.

Regions or countries which may have to take more efforts to conserve land may be less developed. This is evident from the aggregate data analysis. One can design transfer mechanisms to reduce these distributional problems. However many less-developed countries may not have the resources to meet the cost of these transfer mechanisms. The absence of adequate transfer mechanisms to compensate the losses of less-privileged sections could be a major constraint against land restoration.

There is a cost of public resources for land restoration. As we have noted earlier, restoration may have to depend more on public resources due to the positive externality. However the cost of public resources for land restoration include direct and indirect ones. The money that is to be spent for each restoration project is the direct one. However there are other public goods and services to be provided by the state, and hence each dollar spend on one item has an opportunity cost in terms of foregone expenditure on other items. Many less-developed countries have inadequate public investments in much needed services including education and healthcare. These opportunity costs also have to be taken into account in the planning of restoration programs. There may be certain investments in public resources which have a synergy. For example, if human development helps land restoration, and since the former is beneficial to society in multiple other ways, it may be better to invest in it first. Such considerations are also important in prioritising public investments.

The opportunity cost of public resources may reflect in the enforcement of regulations which enable land restoration and also the shaping up of political economy. When governments do not have enough funds, the regulations to use land carefully (say by reducing the excessive use of chemical fertilisers and pesticides or for the rehabilitation of land after the end of mining) may not be enforced properly even if such regulations are in place. Moreover, it will be politically costly to enforce regulations on certain constituencies (say, farmers with small land holdings). We have seen that though Indonesia has put in place regulations to minimise the burning of peatlands, its enforcement is weak due to the cost of public resources and also the political economy pressures from below. On the other hand, these constraints may be less formidable in developed economies.

Environmental Costs

There can also be environmental costs of land restoration in certain cases. This may happen if the strategies of land conservation do not take into account natural ecosystems. For example, tree plantations in grasslands can lead to environmental costs. Similarly, vegetative cover in natural deserts may alter ecosystems. There are areas which may undergo natural degradation, and we may not know the impact of land restoration in such locations. For example, there may be an increase in wind-based erosion or expansion of salty lands over time, but designing appropriate land restoration could be a complicated task here.

Implications of the Cost of Land Restoration

Though we are sure of the benefits of land restoration, there is a need to consider its costs too. This may help policies and practices in multiple ways. We list down a few of these in the following part.

1. The consideration of costs and benefits in land restoration may help us to identify projects and programs which have higher net benefits. These can be taken up on a priority basis. Countries, especially the less-developed ones, may face a scarcity of resources on the one hand, and a large land area which is facing degradation on the other hand, and hence a prioritisation of restoration projects is useful. Assessment of net benefits (after considering the cost) of restoration may help in this regard.
2. For each stretch of land, there can be different levels of restoration. Avoiding soil and water erosion could be one step. Planting indigenous plants and trees could be another level. Allowing it to have natural vegetation is at another level. However the cost and benefits of these different levels of land restoration could be different. Hence the consideration of cost may help in deciding the appropriate level of restoration.
3. Understanding the cost of land restoration may help to design appropriate compensation mechanisms for those who incur losses. This is needed to see that land restoration does not affect certain sections of society. It is also important to see that political economy pressures do not affect the pace of land restoration.
4. Land restoration is a social goal. There are other economic and social goals. There can be a synergy between some of them, for example between human development and land restoration; There can also be trade-offs between some of them. For example, extracting more resources from land for economic development may work against land restoration. It is important to have a socially, environmentally and economically appropriate land restoration. Understanding of costs along with the benefits of restoration is important for this purpose.

To address the challenge of scaling up restoration investment and finances, the government must focus on correcting distorted incentives such as agricultural incentives and replace it with sustainable land use and restoration based incentives. Governments in developing economies may also explore the potential of mobilising climate and development finance and carbon tax revenue by channelling a share of these existing funds towards restoration. Given the wide linkage effects of land across various sectors, coordination of restoration initiatives across ministries may help in streamlining the funds and be more focussed towards the larger mandate of land restoration. Finally, restoration investment can be financed through more innovative partnerships such as collaboration of private and public parties through PPPs, leveraging with philanthropic actors, civil society, crowd funding and multilateral banks.

Chapter 10
Distributional Issues in Land Restoration

All policies and projects affect different sections of people differently and that is true for those aiming at land restoration too. Poor and non-poor may have different needs regarding land, and this may have implications for land degradation and restoration. Usually the less-privileged groups like women and people belonging to specific races, castes, regions and religious groups may encounter additional challenges or their needs may not be taken into account in large-scale programs and policies. These distributional issues require a special attention to see that land restoration is happening at a pace and in a manner which are beneficial to the society at large. Moreover an inadequate consideration of these distributional effects may affect the effectiveness of land restoration programs. These may be affected by distributional struggles as part of political economy.[1] Some of these issues are discussed in this chapter.[2]

People May Have Different Dependencies on Land, and Poorer Ones May Have a Higher Dependence

Economic development need not benefit all equally. There can be those sections who have access to capital and/or education and they move into jobs in service and manufacturing sectors, and their dependence on land (for agriculture and related activities) may come down. On the other hand, those who have challenges in accessing education and urban services may continue with agriculture (including cattle rearing) in less developed or poorer economies. Those who depend on agriculture may get

[1] Though political economy of land is a highly researched topic, we could not see any study on the political economy of land restoration.

[2] There are systematic analysis of distributional impact of specific conservation policies such as Ahlvik, & Kooten (2024). But this chapter takes an overall view focussing the possible negative impact on less-privileged groups.

V. Santhakumar et al., *Land Restoration: A Rational, Grounded and Pragmatic Approach*, https://doi.org/10.1007/978-3-032-26652-1_10

only a relatively lower income (due to the relatively lesser growth of demand for farm products as part of the income growth in society compared to that for industrial goods and services).

This higher dependence on land by the poor is a notable trend in less developed and poorer countries including those in sub-Saharan Africa.[3] It is relatively easier for those who have lesser dependence on land to argue for its restoration as their loss or opportunity costs associated with restoration may be less. One can see such a situation in discussions on wildlife conservation versus farming in human-wildlife conflicts.[4] Governments may favour land and ecosystem restoration policies when these are demanded by powerful sections in society. These may have negative impacts on those who continue to depend on land and ecosystems such as forests. Ideally, there should be adequate compensation to those who incur losses due to government policies. However such a compensation may not be given in reality. This may enhance the ill-fare of those people who are already facing challenges due to their limited economic capability and mobility.

This is not a hypothetical situation as evident from the situation of tribal communities in India. They were traditionally depending on forests for their survival. However forests were nationalised by the British colonial government to extract resources like timber, and through this process the traditional use rights of tribal people over forests were restricted. The post-independent governments also followed the same policy. When forest conservation was tightened due to environmental concerns, severe restrictions were imposed on tribal communities. These had a negative impact on the welfare of these people. Even though a change in Indian constitution was enacted in 2000s to provide a few more rights to these communities over forests, these were not enforced properly. There is a certain visible improvement in the protection of forests in India (Santhakumar 2017). However it was achieved mostly by denying the legitimate rights of its tribal communities. Ideally, the conservation strategies should have taken care of the needs of these communities. This is one case where a particular social group was made to suffer a higher level of losses for the conservation of an ecological resource. This was possible due to the limited power of tribal communities in India since these constitute only a small part of the population in most parts of the country. Moreover they are not politically mobilised in the country (barring in its North-East region).

One may argue that land restoration is useful for poorer people too. Yes, that is true but short-term costs may not encourage them to take the steps towards restoration even though there are long-term gains. This is the relevance of support schemes to reduce their short-term costs. It is good if those who can afford short-term costs and foresee long-term gains can bear a part of the short-term costs of the poor and less-privileged groups in this regard. This can be a rationale of government support for land restoration projects to be carried out by poorer sections of the society.

---

[3] For example, see Barrett et al. (2002).

[4] The need to understand the political-economic structures of this conflict is highlighted in J.D. Margulies, K.K. Karanth (2018).

Better off Sections and Private Capital Degrade the Land Causing Losses to the Poor

The previous section may give an impression that only the poorer people cause land degradation. This is far from true. As noted in the literature, when there is economic growth in poorer countries, there could be an increase in environmental damage and that includes land degradation.[5] The increase in consumption coupled with that of production of not only agricultural but also industrial products which may lead to the discharge of different kinds pollutants in a context where environmental regulations are inadequately enforced may increase the degradation of natural environment including land. This can be harmful to society in general but more for the poorer sections who may be still depending on the nature (land, rivers, etc.,) for their subsistence. However their voice may not be heard due to their limited power in political economy. This demonstrates the need to tighten laws and their enforcement against practices which degrade land, especially those of well off and private capitalists. Such a strategy is good in terms of efficiency, sustainability and also in terms of distributional consequences.

Positive benefits of land restoration for the less-privileged groups may have to be adequately recognised in the design of projects and policies for this purpose. The vegetative covers that are to be re-established as part of the restoration can be those which can yield sustainable benefits to these sections of the population. For example, the needs of fodder and fuel for poorer communities may have to be prioritised in afforestation or tree-planting programs which are aimed at land restoration. The restored land can also be used for activities which can give a higher income to these people, and these may enable them to get out of land-degrading agriculture. For example, eco-tourism which involves wildlife viewing in those lands which are used for grazing by communities in parts of Africa could be an example.

There Are Underprivileged Groups Which Conserve Land

The cultural practices of groups like indigenous people in Latin America help the conservation of land. It is evident from different parts of the world including settlements in Amazon delta. One such account notes the sources of food and income for the people there: 'the food derived from cassava (flour, *tucupi*, tapioca, *croeira* and others), backyards with their fruit trees (papayas, *cupuaçuzeiros*, mango trees and others), small animals (ducks, chickens and pigs), aromatic, medicinal and seasoning herbs (mulberry *catinga*, lemongrass, *capim santo* etc.), the planting of *pripriόca (cyperus articulatus)* and *pataqueira (schizolobium parahyba)* for the sale and production of perfumes by the company, *Nature S. A*, and tourism based on the

[5] This is the idea behind Environment Kuznet Curve hypothesis. For a recent review, Naveed et al. (2022).

potential of the area and the local cuisine'.[6] However, the avoidance of land degradation can be part of the struggle for land by different interest groups. Those who are interested in commercial cultivation including the creation of large-scale ranches, mining, and so on, may be interested in investing in these lands. The owners and beneficiaries of these enterprises may come from urban areas or other countries, and hence there could be a distributional struggle for land leading to its degradation. Such a degradation can affect the livelihood of groups like indigenous people in Amazon.

This requires an economic development strategy that can benefit all sections of society. This may require the conservation of cultivation practices of indigenous people in their territory but allowing extractive activities in other areas. It may also require a sharing of the benefits of these extractive activities with the local people including indigenous groups. There are interesting examples of this kind also from Amazon delta.

There is a perception that modern economic development and globalisation will lead to the decline of the culture of indigenous groups which enabled the conservation of land and forests in their territories. We have discussed this issue in the chapter on communities. However we may note certain enabling aspects of global economic development too. The global appreciation for multiple cultures or livelihoods has gone up. This could be due to the fast spread of information, global travel and due to the cultural and social aspects of globalisation. One can see these clearly in the growth of demand for and the supply of a variety of food all over the world. Food from different nations (Chinese, Indian, Mexican, Italian and so on) are consumed all over the world and this has very little connection with the political and economic power of these countries. The global demand for specific varieties of food has also strengthened the traditions/practices which are specific to ethnicities/regions. These traditions/practices would have disappeared in the absence of such a growing global demand. In general, the 'global demand' for cultural products has increased. There is some evidence that such a demand may go up as part of economic development.[7]

Such an increase in the demand for cultural products may enable many indigenous communities to have a better life by exchanging their traditional products to other people. For example, the demand for Non-Timber Forest Products collected by adivasi/indigenous population from forests are in higher demand (as evident from the increase in prices). The demand for cultural tourism may also go up as part of economic development.[8] It is possible for those sections of adivasi/indigenous population which have access to their traditional resource-base to have a relatively better life through this process. This was seen in India when and where the adivasi people got access to forest resources.[9]

---

[6] https://practiceconnect.azimpremjiuniversity.edu.in/material-production-by-amazon-river-communities-and-its-implications/.

[7] It is noted that there exists an inversely U-shaped impact of economic development on the acceptance of cultural products. See Moon et al. (2016).

[8] This is part of the experience tourism which has a higher elasticity of demand. See Gunter and Smeral (2016).

[9] This was found to be the case where the so called Forest Rights Act was implemented effectively. https://practiceconnect.azimpremjiuniversity.edu.in/a-better-life-for-the-scheduled-tribes/.

Global challenges like the climate change seem to have strengthened these positive aspects of globalisation. The non-destructive use of resources (like forests) by cultural minorities like the indigenous people may offer insights for the sustainable use of ecosystems for the world. It highlights the importance of preserving certain cultural practices even if some of these are not used and adopted by the majority. There could be new information and knowledge that may be generated as part of human progress which may highlight the importance of certain cultural practices in the future, and this merits their conservation to make these available to future generations as possible options.[10]

Women and Land Restoration

Probable connection between women and land are discussed widely in the literature.[11] In certain contexts, women were the prominent users or cultivators of land. Their cultivation practices were more likely with lesser amount of external inputs like chemical fertilisers and excessive use of tilling. These practices are possible only when the fertility of soil is retained. Hence it is in the interest of women to see that the quality of land is not declining. Hence they have an incentive to sustain the quality of land. Moreover in many poorer countries, it is the girls and females who collect firewood, and a destruction of tree cover would mean that they have to walk (and carry firewood for) longer distances. That would increase their drudgery. Moreover, the cultivation of commercial crops (whose outputs are sold in the market for money) is mostly carried out by males. Female members of the household may not have equal access to this money. Hence the commercialisation of agriculture (which in less developed countries can lead to the degradation of land) would also enhance the inequality between men and women in land-dependent societies. In summary, females in general have an interest in preserving the land cover and natural fertility of land.

However women have limited decision-making power over land. In most cases, females may not have the ownership over land. There are societies where land ownership is transferred patri-lineally. Even in matri-leanial communities, actual decision-making or supervisory power may be vested with men. Hence the decisions on how to use land may be taken by men. They may go ahead with changing land-use patterns even when these changes are harmful to women (say through the decline of the firewood in the nearby vicinity). There is also a long standing view that newer technologies in agriculture favoured men (and do not take into account of the needs of women), and this may also aggravate the situation of the latter. Men may not be willing to invest the income from the (commercial) use of land for enhancing the access to services for women (like the installation of stoves which don't need

[10] This is an important rationale for conserving biodiversity.

[11] For a review, see Namubiru-Mwaura (2021).

firewood or water supply at home which mitigates the need to carry water over long distances for domestic purposes).

Hence the differential impacts of the use of land (including practices which may lead to its degradation) on men and women have to be taken into account in projects. There can be a choice between different restoration projects by taking into consideration of these impacts. For example, restricting access to land (which is being restored) may be useful for restoration purpose but that can be costly to women. Hence a restoration effort that can provide biomass intermittently may be better if we take their needs also into account. The cultivation of certain species may be useful for the preservation of biodiversity but whether that will meet the needs of women needs to be considered. The requirements of poorer women may have to be taken into account in agricultural research to see that land-use practices enhances the overall welfare and not only an increase in money income (which may not be accessible to women).

There may be a need to legalise the women ownership of properties. It may be incorrect to consider family head, especially the male, as the de-facto owner of properties. Women may have to be the direct target and beneficiary of the extension programs of agricultural projects. There should be special efforts to include them in training programs on the new cultivation practices.

Women's access to capital and markets may have to be enhanced. An inadequate access to capital and markets for their production may accelerate the transition towards commercial cultivation of mono-crops and worsen land degradation. Microfinance and self-help groups of women may have potential in this regard. These can also enhance the solidarity among women so that they can exercise greater influence in community/social decision-making forums.

In general, the needs of women and other less-privileged groups should be taken into account in land restoration. This is the imperative of human rights. However the consideration of their needs may enhance the efficiency and effectiveness of land restoration projects too.

International Versus Domestic Interests

Land restoration in most cases give benefits to local people. However these may also depend on the reasons which encourage land restoration. If the restoration of land is to ensure its fertility (as a way of sustaining the production of say agricultural goods), then it is more likely to benefit local people. On the other hand, if the ultimate goal of land restoration is the mitigation of climate change and/or the preservation of biodiversity, then these may create more benefits to the global society. Local people even when they are affected by climate change and biodiversity, may not give adequate priority to address these challenges. This can be due to genuine reasons: People from poorer countries may be facing other issues such as poverty and underdevelopment a lot more severely and hence may not use adequate resources towards addressing global challenges like climate change or the erosion of biodiversity. This can lead to distributional issues between international organisations and domestic governments with regard to policies and projects of land restoration.

This shows the importance of analysing the distribution of costs and benefits of land restoration projects for different groups of people, and the design of steps to alleviate losses if any. We have seen in the case of peatland restoration in Indonesia, that a ban on the normal use of peatland can increase the cost for local people (though they may also gain from the global benefits of peatland restoration). The consideration of this is important from a practical point of view (to see that restoration projects get local acceptance and ownership) but also for distributional reasons. (It is the same distributional issue that makes global agreements on actions to address climate change, difficult to achieve.) In summary, there is a need to address this distributional concern to see that land restoration projects are effective and lead to benefits to all regions and sections of people.

The Challenges for the People Who Live in Remote, Vulnerable and Ecologically Fragile Parts

People who live in remote regions including forests or mountainous areas face special challenges, and some of these, and the ways by which people respond to them, may have implications for land. A characteristic feature of these areas is the lack of access to cities and other densely populated areas. We can see that in Amazon delta, Himalayan[12] and Pamir mountains, Borneo island of Indonesia, etc. First, their dependence is higher on the land and connected ecosystems. These ecosystems are fragile and cannot meet the needs of the growing needs of people living there. Moreover there are constraints for the development of non-agricultural activities like manufacturing and services (barring nature-based tourism) in these areas. These can impose limits on the income—and employment-generating activities in these areas. How do people avoid the intensification of their economic activities in these fragile ecosystems, has implications for the land and other ecosystems. Moreover economic development also means demand for a higher level of infrastructure facilities like roads. The construction of these can have negative impacts on the ecosystem including land. Even tourism and the arrival of a large number of outside visitors may necessitate more of such infrastructure, and aggravate the ecological and land degradation. We have seen this leading to frequent landslides and other such events in Himalayas.

There is also another major challenge in these areas. Due to the limited employment opportunities, many youngsters from these communities migrate to cities. Hence there could be a demographic change in these localities. Such a change is

[12] It is estimated that Indian Himalayan region and around has 1,33,06,736 ha land under degraded land (SAC-ISRO 2016). Desertification and Land Degradation Atlas of India (Based on IRS AWiFS data of 2011–13 and 2003–05), Space Applications Centre, Indian Space Research Organisation Department of Space.

also not benign in all these contexts. Migration and hence the decline in the number of young workers may have implications for the continuation of the sustainable land-use activities in these localities. This can also have implications for land and other ecosystems.

It is difficult to visualise a future where all these people are rehabilitated out of these settlements. All these may require a transfer of resources by national governments so that these people do not use land and other resources intensively which can be harmful not only to themselves but also to others. This is another distributional issue within national boundaries.

This is true for all people who live in vulnerable lands. Poverty and lack of opportunities have encouraged people to live in those lands which are vulnerable to one or other disasters. The areas which are prone to flooding or landslides (or even forest fires) are of this kind. The aggravation of climate extremes, caused by climate change, may enhance their challenges drastically. Ideally, these people should be encouraged to leave their lands (and these have to be left in their natural form). However they may not have the resources to move on their own. Moreover their livelihood depends on the activities in the current location. There is a tendency for the state to acquire such ecologically fragile lands.[13] However such a step without adequate compensation may create social conflicts and reduce the effectiveness of restoration efforts. This is a case where societies have to bear the burden of meeting the direct and indirect costs of rehabilitating these people. There is a need to use a part of the public resources for this rehabilitation rather than waiting for mitigating the cost of disasters.

In summary, there are a number of distributional issues in land restoration projects. More often the losers are poorer and vulnerable communities. There is a need to understand their losses and difficulties and to take mitigating measures. This is necessary from a human rights perspective. It is also important to facilitate land restoration effectively.

[13] Such a step was taken in the state of Kerala, India through the THE KERALA FORESTS (VESTING AND MANAGEMENT OF ECOLOGICALLY FRAGILE LANDS) ACT 2003.

Part III
Role of Different Stakeholders in Land Restoration

Chapter 11
Role of Individuals and Behavioural Change for Land Restoration

Land restoration requires actions on the part of individuals and these include farmers, industrialists, public officials, policy-makers, investors and public at large. Why should they take steps towards land restoration? Based on the positive benefits of land restoration that are known globally, and described briefly in the first chapter of this book, one can argue that taking steps towards land restoration is rational on the part of individuals. However, it is incorrect to assume that all or most human beings are rational in all their decisions. The research on behavioural economics has brought out insights into the actual motivations of individuals (Guth et al. 1982; Kahneman et al. 1986 & others). Deviation from rational behaviour is common among individuals based on the insights from behavioural economics, psychology and neurosciences.[1] There are two implications here for land restoration.[2] First, people may take steps towards restoration for rational reasons. Secondly, a set of people may invest in land restoration even if it is not rational, and others may not invest in it even if it is rational to do so. This issue of rationality is different from monetary and non-monetary rewards from land restoration. People can be rational but can put a higher weight to social benefits (and not their personal monetary return) in their decisions. They may take actions towards land restoration even if these are not remunerative to them as investors. One can see a number of altruistic actions of this kind for land restoration.

The role of different motivations on green investments is discussed in the literature. The use of emotions and intuitions are discussed in (Rubaltelli et al. 2010). The approach of behavioural economics is used to develop a taxonomy to define what can be called an 'environmentally sustainable activity' (Beerbaum & Pauschunder 2018). Weber & Johnson (2012) used behavioural economics for the design of a green growth strategy. The use of behavioural economics in the design of energy

---

[1] For a recent review, see World Development Report (2015).

[2] Role of behavioural change for certain aspects which are connected to land restoration is discussed in the literature. Refer https://www.cgiar.org/news-events/news/scaling-land-and-soil-regeneration-through-a-behavioural-science-approach.

V. Santhakumar et al., *Land Restoration: A Rational, Grounded and Pragmatic Approach*, https://doi.org/10.1007/978-3-032-26652-1_11

and climate policy is also discussed (Pollitt & Shaorshadze 2011). The insights of behavioural economics are used to nudge people to make green choices[3] but all these do not deal adequately with land restoration.

This chapter discusses the implications of different motivations of human beings for land restoration and how public policy can respond to such behaviours. There may be a need for policy actions like nudging to encourage people to take steps towards land restoration and a few of these are discussed here. This discussion may also highlight the importance of social fraternities such as faith groups or of role models in encouraging their followers to take steps in this direction.

Self-Interest or Gaining More Income: Motivation for Restoring Land

There are cases of land restoration or specific steps in that direction which may lead to higher incomes. For example, sections of people are willing to pay a higher price for organic products (than that for normal farm products) due to, (a) perceived benefits to themselves from chemical-free food (which is based on a consideration of private health); and/or (b) the benefits of chemical-free-farming to the environment and society in general.[4] When there is such a willingness to pay more, suppliers of organic products may get a higher revenue than what they can get by supplying non-organic farm products. It may be noted that the cost of supply of organic products could be higher (and that may be true for many green or ethical products too), but still, the profit for suppliers of such products could be high. Under such conditions, certain farmers may have an incentive to cater to the organic product market, even though they are not driven by any green considerations. Hence, they may reduce the use of chemical fertilisers and pesticides which may be good for land restoration.

It is rational to reduce losses in future. Farmers may invest in soil erosion for this purpose. People may consider the welfare of their future generations as part of their utility.[5] This too may enable land restoration since such a restored land is needed to provide the goods and serves to future generations.

The willingness to pay for public and environmental goods may go up as part of income growth. Empirical evidence indicates a general positive correlation between income and a willingness to pay for public goods (Bergstrom 1973) and a major share of environmental goods/services is public goods (Kotchen 2012). The higher willingness to pay for the conservation of environment (which is in tune with the self-interest assumption of conventional microeconomics) can also encourage people to make investments for land restoration. They may see the restored land as an

[3] For a review, see Schubert (2017).

[4] For a discussion of organic farming see, http://www.fao.org/organicag/oa-faq/oa-faq5/en/; or articles such as Rodiger & Hamm (2015).

[5] One key concern in economics is whether the welfare of future generations should be discounted or not. See, Arrow (1999).

environmental amenity (instead of or in addition to its role as a resource that yields goods and services). This may lead to higher private investments for the restoration of land. This can be an important motivation for enhancing land cover (including tree cover) in private lands (in addition to, or even without, having a higher income from such land).

However, these motivations may not be enough to conserve the whole land, and there could be a trade-off between the consumption of goods and services, and environmental amenities for many people. There is one argument in the literature that people who come out of poverty or material deprivation may be looking for more material comforts for some more time and they may not necessarily demand experiential goods and services.[6] This can decelerate the growth of income-induced willingness to protect environment (including land) all over the world. However, the income growth of individuals has a certain positive impact on the demand for environmental goods and services and this may help land restoration too.

Other factors, such as education, too can play an important role in enhancing the demand for environmental goods and services (Hirsch 2010; Mobley et al. 2010). All these are compatible with the rationality assumption of conventional microeconomics. Hence, one should expect a certain level of readiness to make green investments including land restoration as part of income growth and educational development in society. This may reflect not only on local environmental goods but also global ones. In that sense, policies which facilitate income growth among a higher share of the population (or an inclusive economic development), may lead to a situation where many more people would be willing to make greener investments including those for land restoration.

Restored land can also be a public good as it provides services like the absorption of carbon dioxide. The provision of public goods (including land restoration in say public or non-private lands) requires coordinated action. In other words, there is a need to address collective action problems for this purpose. This too may indicate the need for a substantial section of society to feel/perceive the need for such goods. Connecting this with the possible increase in the willingness to pay for such goods as part of income growth would mean that a notable share of the population should witness such an income growth and educational development. This too indicates the need for strategies for inclusive economic development in order to facilitate land restoration in a notable manner.

---

[6] Experiencing economic hardship, war or social and political upheavals leads to the development of materialist values (Inglehart 1977).

Yearning for Social Recognition—Opportunities for Land Restoration

People may look for non-monetary rewards from society (Ryan & Doci 2000). These may include social recognitions (such as awards, media exposure) of different kinds. The need for distinction or social status can be used for encouraging green investments (Griskevicius & Tybur 2010) and this is relevant for land restoration too. There can be different kinds of awards and media exposure for those who make such investments. However, the value of such recognition may decline as it becomes less and less distinct—hence, it may be viewed as an effective tool in the early days of green investments. Its usefulness may decline when such investments become more common (due to all kinds of motivation described in this article). There is a general perception of the inferiority of extrinsic motivation and the superiority of intrinsic ones (Ryan & Doci 2000), and the latter is considered in the next section.

Intrinsic Motivation and Land Restoration

People are motivated not only by financial (material) and extrinsic (like recognitions) incentives, but also by an intrinsic motivation (Ryan & Doci 2000). This motivation may dominate in some individuals. Experimental or quasi-experimental settings and the presence of anonymous giving for social and environmental purposes have demonstrated the influence of such an intrinsic motivation. Such a motivation to 'do right' is somewhat different from the motivation to take actions due to the concern about others. Hence, there can be investments for land restoration driven by intrinsic motivation even if it is not monetarily rewarding and will not bring in any extrinsic/social recognition. We have seen a number of environmental activists who are involved in different aspects of land restoration by sacrificing material aspects of life and without any concern about social recognition. However, they seem deriving a lot of joy and leading meaning life through their actions.[7]

What we have considered so far are rational reasons which may encourage people to pursue land restoration. However, it is incorrect to assume that all people take rational decisions in all contexts. The implications of not so rational decisions for land restoration is discussed in the following part.

[7] One such group of people that we have noted is Navadarshanam in Tamil Nadu in India. Refer https://navadarshanam.org/.

Not So Rational Actions and Land Restoration

Behavioural economics note that people may have certain inertia or status quo bias which may discourage them from taking actions even if they know that such actions are beneficial to them. Such an inertia would mean that people need not invest in green or ethical projects like land restoration even if they think that these are good to them and have the resources to make such investments. One example could be the laxness in replanting of saplings of trees after the matured ones are harvested. The other example could be the inadequate strengthening of soil and water conservation structures after these have weakened. Such a trend can be noted in private lands.

This tendency underscores the importance of the way policies are framed or announced or the need to design appropriate framing and announcements. For example, a project which is found to be beneficial to all people in a set can be offered in such a way that the default option is to be ‘in’ and it can be made the responsibility of those who do not want to be part of it to ‘opt out’. These are called decision defaults (Benartzi & Thaler 2004). This may probably be better than giving the choice to ‘opt in’ under certain situations. Hence, it may be useful to mandate certain investments such as tree-planting or soil conservation rather than giving the option of whether to do so or not, to specific land holders.

There Could Be Myopic Decision-Making

There can be, what is called, cognitive myopia (Benartzi & Thaler 1995; Read et al. 1999) which may encourage people to neglect gains in the longer term and weigh immediate losses heavily. This is related to the use of inappropriate discount rates (Weber et al. 2007). This habit may work against green investments including those for land restoration whose returns are derived in the long term. This could be a severe problem if poverty encourages people to discount future heavily. This can be a major constraint against land restoration in those lands which are used by poorer people.

The mitigation of this problem may require a combination of investments which yield returns in both medium-and long term. Or land restoration projects have to be redesigned in a way to generate both short-term and long-term results. It can also be attempted by monetising the long-term gains which would ensure financial returns to land holders in the short-and long-term. Those who are exposed to medium-term returns may gradually accept those opportunities which yield longer-term returns. The provision of information or feedback intermittently on how one’s investments contribute to green goals may also help in overcoming this constraint to some extent.

There Could Be a Greater Focus on Avoiding Losses Rather Than Enhancing Gains

Behavioural economics has demonstrated that people may value losses and gains (from a given point or benchmark) differently. The marginal disutility associated with the loss could be higher than that related to the gain. This may have some implications for land restoration or other green investments if these results in a reduction in financial gains immediately (Weber & Johnson 2012). There could be a role for reframing upfront costs not as losses but as foregone gains (Podesta et al. 2008). It could be that people may not be willing to divert their accumulated endowments for purposes such as land restoration. On the other hand, there could be a higher willingness to allocate their potential future gains into portfolios of such investments. The tightening of environmental regulations and the consequent signalling of the potential losses from environmentally harmful investments could facilitate such a change (since this may spur green investments due to the motivation to minimise the losses, an issue discussed earlier).

People Are Willing to Sacrifice for Their Notion of Fairness

Experimental studies note that people are willing to sacrifice based on their notion of fairness (Fehr & Henrich 2003). Will that notion of fairness uphold towards nature is a contestable issue. It may be so for a set of individuals, and then they may be willing to give the due share to nature. This is not unusual among environmental activists as they may often use metaphors of nature as a living being (nature has life). It is not unusual to see literary mentions of land as the mother. Hence, social norms or ideas or beliefs (and these may come from religions) that encourage people to view aspects of nature (including land) as living beings, may help in this regard.

Actions Driven by Instincts and Peers

Against the expectation of rational behaviour on the part of individuals, people may not act after deliberation in reality. Instead, they may act instinctively (Word Development Report 2015). Hence, the possible benefits of specific actions may not be considered before taking actions. This would mean that even if land restoration is beneficial to oneself, there may not be adequate action in that direction. Another aspect is that the actual behaviour of people is influenced by what their peers do. This is also without much deliberation as noticed in the literature. This may create multiple equilibriums in terms of land restoration. If people in general are concerned about avoiding land degradation and take steps to minimise it, many others may also take steps in the same direction (even without thinking much about the benefits of such restoration). On the other hand, people may not take actions towards restoration if their peers are not doing it (irrespective of the benefits and costs of such actions for specific individuals). This insight may be useful to design actions which are directed at specific peers. There are neighbourhoods and other groups of people who come together through some identity or faith or ideology and changing orientations of the group as a whole towards land restoration may be effective. This can be a rationale for using faith groups or religious communities to facilitate actions towards land restoration.

Actions Based on Incorrect Models of Mental Reality

The third aspect of the actual behaviour is that people act on the basis of mental models of reality but some of these models may not be grounded in actual reality. There can be such models on the functioning of natural and social systems in the minds of people and some of these models could be unrealistic or inaccurate but people may behave according to such models. Here too there are possibilities of multiple equilibrium. There can be a set of (not necessarily correct) models of reality which may facilitate the conservation of land and such assets. For example, there may be groups which consider the 'nature' as divine, and its degradation as 'sin', and they may take steps towards its conservation (whether such a belief system is correct or not). But there can be several other models which may encourage people to take least care of land. This insight also shows the importance of creating mental models of reality which can be rational and which highlights the importance of land restoration. Land restoration may not happen due to research insights on its importance, policy documents which outlines actions for this purpose or laws which are created by legislatures. It has to be part of the thought process of people at large. This indicates the importance of widespread communication and popular discussions on the problem of land degradation and the need for its restoration.

Non-internalisation of Relevant Information

It is also incorrect to assume that people internalise all relevant information and make corrections in their choices. Even when people encounter new information, there can be a filtering: that which conforms to their pre-judgements may be accepted and others may be rejected. Even educated people in the US who are followers of the Republican Party may not accept that that climate change is a serious environmental problem (despite the availability of research insights and data in public domain). Hence, peer-pressure or changes in the beliefs and attitudes of groups may be a lot more important in ensuring large-scale restoration.

Relevance of Social Norms and Trust

The role of social norms cannot be neglected in this regard. We discuss the role of community norms in another chapter, where we see that the hold of these norms may weaken over time when the mutual dependence among community members declines. However, this does not mean that norms can be neglected, and people's behaviour can be regulated merely through laws and their enforcement. Socially desirable actions and behaviour should become part of social norms. This may encourage people to follow these to avoid social/peer-group disapproval. On the other hand, if people behave appropriately only due to the fear of law enforcement, such behaviour may not be forthcoming if the enforcement is weaker in certain contexts. Moreover, ensuring perfect law enforcement can be very costly to any society. Hence, social norms, if these encourage people to behave appropriately even without law enforcement, are efficient.

Trust also plays an important role in society. Trust is based on a belief that the other would behave appropriately even if circumstances allow that person to behave differently. This happens when people 'do the right thing' irrespective of the circumstances, and the society in general expects most people to do so.

How do desirable social norms and trust develop in societies? This is a challenge in less-developed societies where community norms may have weakened due to increasing dependence on wider markets for goods and services, capital and labour. In the place of these community norms, social norms which are valued across communities have to develop. This may have to be based on what is what is called a 'generalised social trust'. An inclusive economic and human development may be playing a role in the development of such a trust. This may be indicated by the higher prevalence of such a trust in countries like those in Scandinavia. In that sense, an inclusive economic and human development, that we have discussed as important to enhance the incentives for land restoration in a previous chapter, may be important for the development appropriate social norms and a generalised social trust which can also contribute to such restoration.

Conclusion

Though land restoration is a rational investment by considering the individual and social benefits, we should not presume that all people will take steps towards such restoration through rational deliberations. In certain cases, irrational behaviour (based on the way the rationality is defined in economics) may lead to land restoration. A part of environmental activism can be of this type since a set of activists may be following a mental model of reality which may not be closer to the reality (as evident from a closer assessment of evidence). People may follow social norms or intrinsic values and these may enable restoration in certain contexts. However, in most cases, irrational behaviour may lead to land degradation (even though it is rational to restore land for individual and social benefits). There can be myopic behaviour which may lead to underestimation of future losses (and inadequate efforts towards land restoration may lead to such losses). There can also be a status quo bias or inertia which may discourage people from taking appropriate actions even when they know that these are beneficial to themselves.

This possibility of irrational behaviour implies that information on the private social benefits per se may not be enough to encourage people to adopt restoration measures. The use of peer-pressure, widespread communication and also the creation of appropriate social norms are important for land restoration.

Chapter 12
The Role of Communities in Land Restoration

There are forests, grazing lands, mangroves and wetlands, peatlands, deserts, etc., which are still controlled by communities (or people living in a location as a group). These entities may not have private property rights but they may have certain use rights sanctioned by tradition or historical patterns. Though there is an expansion of private-or state-ownership to these community-controlled lands, these exist a significant part of the overall land in many parts of the world under community control. There could be a higher level of degradation in these lands as part of socioeconomic development. This can be partly due to the change in population size, and an increase in economic activities which may encourage intensive use of these lands. In certain cases, the abandoning of these lands can also lead to the degradation (due to the absence of care and management). How do we stop the degradation of these community lands has to be an important concern for land restoration globally. Certain aspects of the evolution of community management of land and its implications for land restoration are discussed in this chapter.

The role of communities in land restoration is discussed in the literature (for example, Sze et al. 2022; Dai et al. 2024; Tewodores et al. 2025). However, we take a basic approach to communicate the strength and limitations of communities in this regard in this chapter.

Traditional View of Economists—'Tragedy of Commons'

The initial view of these community or common lands among scientists and economists was one of pessimism. These lands are not owned by any one individual but are used by a number of people, and hence, the lack of incentive of any one to use the resource in a sustainable manner, was noted in the literature as the 'Tragedy of Commons'. It is a situation where individuals whose actions are driven by self-interest, cause the depletion of a shared resource. Since the individual benefits

V. Santhakumar et al., *Land Restoration: A Rational, Grounded and Pragmatic Approach*, https://doi.org/10.1007/978-3-032-26652-1_12

directly from using the resource, while costs of overconsumption are shared by all, it creates an incentive to exploit if she is driven purely by self-interest. Overfishing is an example—each individual fisherman aims to catch as many as fish as possible for increasing their profits. But if every fisherman in the local area behave in a similar manner, fish population would decline. Similarly, the overuse of communal lands by a group of people can be harmful both in terms of the surplus creation and the sustainability. By following this analysis, there was an argument for either privatising common lands or the use of state ownership and control.

Empirical Evidence (by Elinor Ostrom and Others) on Sustainable Management of Commons by Communities

However, there were a number of empirical studies after the formulation of the 'Tragedy of Commons' on the management of common properties. Elinor Ostrom, a political scientist from the USA was at the forefront of these studies, and she received Nobel Prize in economics for this work later on. These studies, which were conducted in less-developed or poorer countries like Nepal, demonstrate that most communities were managing common properties (forests, water bodies, irrigation systems) without any major signs of deterioration of the quality. The relationship between the members of these communities and the norms which they could use to regulate the activities of each member were found to be adequate to ensure the sustainable use of these resources in most cases. When a member is violating these norms and doing something against the sustainable management of community resource, others may punish the person, or other members who may continue to interact with him/her. Such a punishment can include social ostracisation of one or other kind and is based upon norms and practice than legal backing. This punishment or threat of punishment by the community itself was enough to deter members from deviating from community norms. There may be specific norms (institutions) mediating the use of resources in each context.

Theoretical Understanding on What May Enable Communities to Manage Commons Sustainably

Economists have used game theory to model the repetitive interactions between community members and its impact on the management of community resources including land. This is called folk theorem in economics. They could see that it is possible to have an equilibrium wherein one or other punishment strategy may encourage the members to follow community norms which aim at the sustainable management of common properties under certain conditions. These punishment strategies could be many: not interacting with a person who violate norms forever or for a fixed number of terms afterwards, or public shaming by displaying names of the violators on a common community board. These punishment strategies may make a person's gains from one-time violation of community norms lesser than the potential losses due to the punishment by others afterwards. However, such sustainable management of common property resource by communities, without any legal intervention is possible only under certain conditions. These conditions are discussed in the following paragraphs.

Mutual Dependence of People Within Communities

Theoretically, these norms may be followed if the interactions are infinitely repeated. Though human beings are mortal, families who live in rural areas and have been traditionally depending on common lands may foresee their future generations also interacting among each other. This situation may be closer to infinite interactions. The problem with the finite number of interactions is the following: If they know that interaction is going to end after a finite number of rounds (say 10 rounds), they have no incentive to cooperate at the tenth round. This is known to all. Due to this, they may stop cooperating at the ninth round also. Since this is also known to them, they may not cooperate at the eighth round and so on. Hence, cooperation need not be the equilibrium outcome if they know that their interactions are going to end after a finite number of rounds. Families living in a location for generations and using a specific community resource may be reckoned as those who expect to interact infinitely, and may meet the requirement of this condition for cooperation over time. Alternatively, infinite repetition in the real world may also be interpreted as not knowing when the game ends. This assumption is more realistic than it seems—in the case of community managed land it implies members interacting amongst each other without knowing when these interactions will cease to exist. The cases which are studied by Eliner Ostrom and others may be of this kind.

Not High Discount Rate

The second condition for the sustenance of community norms is that no individual should discount his/her future gains heavily. If it is done, the value of future gains may become very small and lesser than the current gains, which may encourage the member to violate community norm (even if there is a punishment by other afterwards). People who face extreme poverty or one or other severe vulnerability may use a very high discount rate (since meeting the current needs become too important for them and they cannot plan for the future). There may be situations where the socioeconomic reality may encourage one or other member to violate community norm so as to make a substantial gain, even though the person is aware of the loss due to the non-cooperation of others in future. When a number of members violate these norms, observance of norms and sustainable management cannot be an equilibrium outcome.

Reality—Socioeconomic Context is Changing

Though community norms could ensure the sustainable management of common lands as observed in the empirical studies by Elinor Ostrom and others, and explained by the folk theorem, social economic reality is changing over time, and that conditions which are required for the effectiveness of community norms may become less and less prevail over time. Let us consider a few of these possible changes.

Declining Mutual Dependence Among Members of Community

There can be different reasons for the decline in the mutual dependence among members of communities. There can be a spread of education and with it, younger members may move out of agriculture and related activities. They may migrate to cities or other countries. We can see this happening in interior parts of central Asia where a significant share of youngsters migrate to far away places including other countries. One can see this dependence on wider markets not only in the case of labour, but also in the case of capital, goods and services. All these may reduce the dependence of community members on each other. In certain other contexts, one can see immigration of outsiders also. For example, in many coastal areas of India, small scale fishing which was carried out by traditional fishing communities, currently depends on migrant workers from far away places. These new workers are not aware of and are not bound by traditional community norms of fishing communities. This can change the composition of the community, thereby diluting the social norms and practices and consequently have a negative impact on the management of common lands.

State-Management as an Alternative to Community Management But There May be Weak Enforcement

When communities fail to manage common lands sustainably, the theory and empirical reality suggest that the state should play a greater role in regulating the use of resources. This could be in the form of instituting state ownership and its enforcement. It can also be through the regulation of the use of resources. For example, government in collaboration with communities have introduced rules regarding the use of grazing lands in the mountains of Kyrgyzstan. Community forests have become public forests in many parts of the world through this process. The territorial seas where the fishing is regulated by communities have come under the control of governments which in turn use different regulatory measures to avoid overfishing. The expectation is that the state would enforce its ownership rights, but there are cases where the lack of adequate ownership leads to a situation where public ownership would mean 'no ownership'. The cost of enforcement could be higher and the governments especially in less-developed countries may not have enough public resources for this purpose. Given the declining control of communities, individuals may use resources in unsustainable ways by using the weakness of public enforcement.

Private Sector May Come in But Its Impact Could Be Harmful in Contexts When Governments are Weak in Regulation

One solution that is projected when communities fail to manage common properties is the privatisation or the institution of private property rights. This is theoretically right if the sole purpose is the enhancement of the efficiency of the use of common property resource. For example, fishery in a lake can be used efficiently (without overfishing and depletion of fishing stock) if there is private property right. However, there are cases where the sustainable use of common property may have positive externalities. The sustainable use or restoration of common or community lands is an example. Private owners may not consider these positive externalities adequately. There may be a need for public regulation. However, the limitations of public enforcement, especially in poorer countries, may work against the adequate consideration of these externalities in the use of these resources. There may be a need for non-governmental organisations to compensate for the weak enforcement of community norms and state ownership.

We consider a few examples of community management with the collaboration of other stake-holders in the following sections. These are based on case studies carried out by Santhakumar and his colleagues as part of his work at Azim Premji University.

Case I: An NGO Enabling Community Management of Land in India

The Foundation for Ecological Security (FES) has encouraged people to conserve community lands in certain semi-drier localities in India. The localities that it has selected are the ones where people continue to have a strong interest in protecting common properties, such as water or uncultivated lands. First, most of the population in the area or the selected villages continue to be involved in agriculture. Jobs in industry or services have not taken many people away from these localities. These are also drier areas with limited rainfall. The preservation and careful use of a limited, available water resource is important for the continuation of cultivation there. Another important source of livelihood is cattle rearing (small ruminants, like sheep or goats). The fodder for these animals is met from the non-private and uncultivated lands. That would give them an incentive to sustain biomass in these lands. Forest fires which lead to the destruction of this biomass can affect their livelihood, and hence, they have an incentive to take collective action to prevent forest fire.

Though there is seasonal migration of people from these localities, there is not much permanent immigration to or out-migration from these localities. Hence, the community is broadly homogenous with income opportunities available through common activity—this leads to the persistence of a reasonably higher level of mutual dependence among the people there. All these may serve as enabling factors for the involvement of people in the management of common properties.

However, this is not to argue that the people in these communities would have taken steps to conserve and manage commons even without the intervention of an external agency, like FES. This is evident from the fact that there was a certain disarray in the management, and hence, there was a degradation of these properties before the intervention of FES in these localities. We may not be able to trace the exact reasons for the inadequate success of the collective action (especially, when we consider the prevalence of enabling conditions that we have discussed here) before the intervention of an external agency. There could be certain other factors that might have discouraged them from taking collective actions to protect common properties adequately. Rights to these resources could be an important issue, especially, when communities do not have secure rights in, what is seen as, state property. There could be a certain lack of information on the need to protect these resources. Given the poverty and vulnerability of a substantial share of people in such localities in India, they may be concerned about other pressing needs and that may lead to the neglect of certain common property resources. (This could be equivalent to the use of higher discount rates as discussed in economics literature). Alternatively, people may not, by themselves, take certain actions which are beneficial to them and that may require 'nudging' by others (an insight from behavioural economics).

The future sustainability of the community management of resources, in these areas, may depend on a number of factors. The possible trends in this regard can be noted from what is already happening there. For example, the enhanced use of LPG as a cooking fuel has reduced the local dependence on fuelwood, and on uncultivated common lands for this purpose. This is beneficial in the sense that they will not use common land for this purpose. However, such a reduced dependence can also reduce the incentive to take care of common properties. This may be true for a number of other livelihood activities. Currently, the grazing by sheep (or goats) and the consequent requirement for fodder is an important need for people, and this encouraged them to use uncultivated common land on a sustainable basis. There can be a change in this need. The reduction in the use of water in agriculture or the availability of other sources (such as canal water from a distant source) can also lead to the disuse or the lack of interest on the part of the community in the proper maintenance of local water resources. However, such a change need not happen if the commons are seen not only for their commodity—or use-value but also for its social-cultural values. Even with changing nature and extents of dependence on commons (for its economic benefit), these are known to be an important part of the social-cultural ethos of many of these communities and are acknowledged for their ecological importance.

Case II: India's Experience of Tree Cooperatives

When conventional community management fails to manage land cover well, and when privatisation and state control may not be appropriate, there will be a search for alternative organisations. Cooperatives are an important category in this regard. Cooperatives of members or workers manage business enterprises in different parts of the world. There are successful cooperatives in India in the domains of milk processing and distribution. Based on this experience, there was an attempt to create cooperatives and a national federation for cultivating and harvesting (sustainably) trees in India with the aim of enhancing tree cover in degraded lands. However, the experience in this regard is not encouraging. This may be due to the fact that cooperation for the planting of trees (and the generation of income and/or other outputs) did not yield adequate gains to encourage people to take the efforts towards sustainable collaboration.

Case III: Territories Controlled by Marginalised Groups (Like Quilombos) in Brazil

Community-land ownership exists in Brazil in the territories where indigenous people and Quilombolas live. Some of these communities have control over larger territories, and most of these areas sustain secondary or even natural forests. Moreover their use of land may not be that intensive like commercial farming or ranching. The leasing in of land by outsiders is made illegal in indigenous territories based on a Brazilian law (amended in 1988). It has been noted that though there is deforestation in Brazil, indigenous territories are largely unaffected by it, especially in Amazon. These are enablers for land restoration.

However, these territories are also facing drought and other such challenges. Moreover, these people (especially Quilombolas) need sufficient incomes to mitigate their poverty and vulnerability, and hence, there is a need to explore income-generating options without degrading land. Though activities, such as ecotourism or cultivation/ collection/ processing of land-based products and their marketing can enhance incomes, there is a need for higher support (in terms of capacity-building and capital investments) for this purpose. Moreover the community ownership/control over territories also make collective action a lot more difficult. There is a need to strengthen those initiatives which aim at addressing these challenges.

Case IV: Territories Controlled by Indigenous Groups (Like Dayak) in Borneo, Indonesia

Dayaks are the indigenous people of the Borneo Island of Indonesia.[1] Sections of them continue to depend on forests for their livelihood. There are different non-timber forest products which are collected and sold by local people. Sustainable harvest of bamboo and rattan can be a source of income and livelihood for them. Timber, if it is harvested sustainably and carefully, can meet their housing requirements and be a major source of income for local people. As an alternate income-generating activity, small-tower-like structures are constructed near houses to attract birds, and their nests are collected and sold for export to Chinese markets. Hence, there are different ways to meet the needs of cash income for these forest-dependent people. There are other products from forests like Illipe nuts and honey which can also be harvested without destroying forests. These can generate a higher income for these people if these are processed and marketed (probably with the help of external agencies).

Forests also meet the subsistence needs of people. People can practice slash-and-burn cultivation with adequate fallow period in between. This is the way they cultivate paddy, which is the staple food for these communities. They also cultivate cassava in small plots without destroying forests. There is a variety of fruits (including durian, jackfruit, etc.); water is cleaner (if managed well), and some of these villages also produce hydropower, locally.

The lure of more money is not the sole incentive of these communities. The communities live cohesively and life in these villages with their cultural attributes is meaningful for these communities. It is possible for one member of the community to take actions that may destroy forests, but community pressure is adequate so far to prevent such actions. Hence, the value attached to community life and culture is also enabling the protection of forests.

The life of a few villagers in the community also gives the impression that people there can have a decent life even without abandoning forests and cultural practices. Indigenous people do not seem poor here when we compare their situation with that in other countries. They have access to basic food, housing, reasonably clean (cleanable) water, toilets, etc. Food is diverse enough with adequate amounts of protein, vegetables, leafy greens and starch. People interact with each other regularly and have enough community platforms for socialisation. They are not deprived of the benefits of modern education (with schools in villages set up by the Government of Indonesia). Basic healthcare is available, though more advanced medical treatment

[1] https://practiceconnect.azimpremjiuniversity.edu.in/private-businesses-in-forest-conservation-the-case-of-forestwise-in-borneo-island/.

may require transportation to cities. Though youngsters go to cities for higher education, they continue to have close relationships with their families. The situation in Borneo is closer to that in indigenous territories in Brazil and other Latin American countries.

Case V: An Example of Wildlife Tourism in Collaboration with Communities from Africa

Phinda private game reserve[2] which is located in KwaZulu-Natal, South Africa, is a private park attracting visitors for viewing wildlife in natural surrounding. It has different ecosystems. Though it was started by one individual by using his 170 sq. km, agreements with local communities could expand its size to 286 sq. km. Members of local communities take up the job of workers in the park, rather than cultivating their land (and such cultivation would have reduced the effective size of the park and created animal-human conflicts). We could interact with a set of workers, and they and there is a general understanding that the income for them from the park is higher than that from cultivating their less fertile land. Moreover such cultivation can lead to land degradation. There can be such cases of partnership between communities and private firms which can avoid land degradation but without leading to a decline of the income of community members.

---

[2] https://www.andbeyond.com/destinations/africa/south-africa/kwazulu-natal/phinda-private-game-reserve/?utm_source=google&utm_medium=organic&utm_campaign=organic-global-mofu-gmb-listing-phinda-private-game-reserve.

Chapter 13
Role of Non-Governmental Organisations

Many social or public activities may require the intervention of non-governmental and civil society organisations for different reasons. First, government organisations, though have the mandate, may not be responsive enough to understand and respond to all social needs. NGOs and CSOs can play an important role in pressurising governments to act. Secondly, the government is not so capable to carryout innovations in social actions. NGOS and CSO can have a pioneer role in this regard, and ideally, the government can learn from their innovations to scale up actions. All these are relevant for land restoration too. As part of the work of the G20 GLI, we have seen a variety of NGOs working on one or other aspects of land restoration or avoiding land degradation. This chapter discusses the advantages and limitations of NGOs and CSOs in this regard, and how to make their interventions a lot more useful in enhancing the scale of land restoration.

The role of NGOs in land restoration is assessed in specific cases by researchers (for example, Harada et al. 2022), but we take an overall view of the possibilities and challenges of these organisations in this regard here.

A Wide Variety of NGOs

There are a variety of NGOs and CSOs which are active in land restoration. First is those by environmental activists or those concerned about the sustainability of conventional path of economic development. They are more likely to be influenced by one or other ideological version of environmentalism and this may influence their actions of land restoration too. We have seen individuals who have almost left normal economic activities and pursue alternative paths of life and land restoration is seen as part of such an alternative. Navadarshanam[1] is one example. There are

[1] https://navadarshanam.org/.

V. Santhakumar et al., *Land Restoration: A Rational, Grounded and Pragmatic Approach*, https://doi.org/10.1007/978-3-032-26652-1_13

individuals who use their own wealth, form charitable organisations, and buy (barren or degraded) land and allow it to be restored. There are also NGOs which do a variety of activities including land restoration by taking money from governmental or international organisations or philanthropic foundations.

NGOs which are concerned about the deterioration of natural environment have taken efforts to bring the agenda of conservation to the forefront. The governments may respond to the demands of these organisations. This is true for land restoration too. They have been at the forefront in highlighting degradation carried out by powerful groups like commercial plantations or industrial polluters including mining industries. A few of them have also tried out constructive work in land restoration (as noted in the previous session). However, there are newer players emerging in this domain. One can see corporate social responsibility organisations becoming active in land restoration.[2] Of late, impact investors or organisations which mobilise them also get into land restoration. Their purpose is not profit-maximisation and hence their actions have to be seen different from that of private firms which are involved in the business of land restoration.

Different Kinds of Actions

These organisations carry out different kinds of actions towards land restoration and some of these mentioned in the following sections.

Awareness Creation

Almost all these NGOs or CSOs which are involved in land restoration contribute directly or indirectly to the creation of awareness on the importance of restoration or avoiding land degradation. These may include highlighting the problems of conventional land-uses (commercial agriculture, mono-crop forestry, industrial pollution, indiscriminate mining without appropriate rehabilitation of land, etc.). This is a lot more useful in less-developed and poorer countries where people's access to higher education (which may deal with the importance of environmental conservation) is limited. Moreover, reading materials or exposure to environment-oriented audio-visual materials from different sources for self-education are limited in these contexts. It is here that the personal and direct communication by NGOs matter.

[2] One such case from India is documented here: https://practiceconnect.azimpremjiuniversity.edu.in/ecological-assessment-of-a-biodiversity-park-status-and-future-challenges/.

If we take regions where environmental awareness is at a higher level, the role of NGOs in creating that awareness is important at the beginning.[3] In fact, the awareness of government officials and media personnel could be limited, and NGOs may play an important role in compelling them to have a higher level of knowledge on environment-related matters.

Campaigns Against Polluters and Those Who Destroy Natural Environment

NGOs are at the forefront of adversarial actions or agitations against firms which degrade environment including land. One can cite a large number of such cases from different parts of the world. People at large may not be aware of the actions and impacts of these firms. Even when a few people are aware, there could be difficulties for them to take an interest in public affairs. Powerful firms may collude with government officials and politicians to hide their impacts on environment. Or there can be powerful interest groups behind such environmental destruction. Hence, one or other individual may not be able to organise agitations. It is in this context that environmental activists and organisations play an important role. Even when there are laws against environmental destruction, these may not be enforced adequately in less-developed countries, and such an enforcement happens only when there is a demand from the common masses. Environmental NGOs play an important role in articulating this demand.

Lobbying to Change Policies

Policies do not change that often. There is a cost in changing policies. Moreover, substantial time of the legislature may have to be used for this purpose. Hence, governments may continue with past policies if there is no adequate external pressure. Though lobbying is an important way to encourage the government to change policies, usually this happen to favour specific interest groups (like industrialists). Issues which affect people at large may not be part of lobbying as they may not come together and there could be coordination problems. Environmental issues are of this kind. However, NGOs may play an important role to address the coordination problems in the case of environmental issues which affect a large number of people. The ideological interest of these NGOs may encourage them to sacrifice personal gains to campaign for policy changes. One can see such a role of NGOs at the international, national and local levels. This is important for land restoration too.

---

[3] This is evident from the Indian state of Kerala where NGOs like Kerala Sasthra Sahitya Parishad (KSSP) played an important role in creating environmental awareness.

Actions Adversarial to Governments

Governments, though are responsible to correct market failures (and land degradation is the reflection of such failures), there are government failures of different kind. First, it may not correct many market failures. Even when it institutes laws to regulate market failures, these may not be enforced properly. These need not be due to inactiveness, ineffectiveness or inefficiency on the part of governments. This can be due to the political economy pressures. Those who gain from degrading land could be economically powerful. Those who are affected by it may not be electorally powerful. The section of society which may be affected by land degradation (say future generations) may not be present currently. Hence, inactions or inadequate actions on the part of governments towards land restoration are common.

It is in this context, that NGOs may carryout adversarial actions like protests and agitations to activate governments. These activists behind these agitations may foresee losses associated with degradation and may be able to go beyond personal interests in this regard. One can see a number of local, national and international actions of NGOs for this purpose.

Reconstructive Actions

In addition to adversarial actions against the government, NGOs also take up reconstructive work. Land restoration is not an exception. There are a number of locations where an NGO restores land by mobilising altruistic resources (which are not seeking profits). They may mobilise money from individuals or corporate social responsibility arms of private companies. Voluntary labour may also be used for this purpose. Such projects include regeneration of barren or degraded land. Due to the close attention of individuals behind such projects, these may be a lot more effective in terms of regeneration rather than large-scale projects carried out by national governments or international organisations. These projects can be cost-effective also due to these reasons, and the paid-cost could be lower due to the voluntary labour provided by individuals. Some of these projects may aim at generating certain financial resources/gains through restoration but others may aim at only environmental or aesthetic benefits. A few of these projects attempt to incorporate the interests of local people whose life is affected by degradation or restoration.

Innovative/Pilot Experiments

Most of the reconstructive projects carried out by NGOs may be using different kinds of innovations or can be seen as pilot experiments. They can be innovative in terms of the diversity of vegetative cover and the way of generating resources from restored land, mobilising people and resources for restoration projects, etc. There can also be newer techniques of soil and water conservation, propagation of indigenous species, plant support mechanisms and so on. NGOs may have an advantage in terms of social innovation compared to governmental and international organisations. The ideas with drive these NGOs may also encourage them to innovate in ways which cannot be accomplished by government or private companies.

Grounded Research and Information Collection

NGOs in certain cases help bring out ground level information on degradation. This may be through micro research and/or data collection. This is useful in countries where governmental data-collection systems are weaker, or in contexts where micro issues of degradation are neglected in such systems. This information may be brought to the attention of media and may become part of public discourses. Hence, there can be a lot more communication to local stakeholders which may facilitate their participation on public debates (compared to the data collection and research by governmental organisations). In certain cases, private/corporate or governmental organisations may have an interest to hide the extent of degradation, and this can be brought to light through the actions of NGOs and activists. Public research organisations may enter only later in researching issues over which there is a higher level of conflicts or contestations.

NGOs as Partners of Large-Scale Projects

There are NGOS which become implementers of aspects of large-scale projects. Their role could be to involve local people, improve their livelihood and so on. Governments or international organisations may not have the reach or the connection with all stakeholders, especially marginalised social groups. These may not have the capacity to understand and respond to the needs of ordinary people. NGOs may have advantages in this regard. We have seen local NGOs/community groups working as partners of the Government's peatland restoration agency in Indonesia. However, there are problems in this partnership with NGOs. Their actions may be influenced by the resources available through these projects. There can be financial incentives which drive these organisations and these need not always be in tune with the goals of effective land restoration. Moreover, such NGOs may be seen as coopted by the large power structures and this can affect their credibility. However, the participation of NGOs in the implementation of large-scale public projects is a growing trend and land restoration is not an exception.

Share-Holder Activism in Corporate Organisations

A newer kind of activism has emerged in corporate affairs and this has implications for controlling environmental degradation. When activists highlight the environmental degradation by private/corporate companies, these can be neglected by their management. However, a set of (new generation) NGOs, attempt to buy the shares of large publicly listed companies by mobilising resources from individuals who are concerned about environmental degradation. Then these organisations may send their representatives to the shareholder-meetings of these companies. They may be able to highlight the environmental excesses of these companies and can argue for changes in corporate policies. Visible presence of these organisations and their voicing of environmental demands may attract public attention and hence cannot not be neglected by the corporate leadership. There are investors who have shares of publicly listed companies but are concerned about environmental issues, and their collective action to push for changes in corporate policies can be coordinated by this newer kind of activism by NGOs.

Possibilities and Limitations of NGOs in Land Restoration

There are certain benefits and limitations of NGOs action in land and ecosystem restoration and these are discussed in following paragraphs.

Possibilities

NGO's actions make people at large, elected representatives, and government officials aware of the importance of conserving environmental resources including land. Cases are documented where all political parties unanimously supported destroying parts of forests or land for the so called development projects, but the agitations of NGOs encouraged the former to rethink their projects finally leading to the conservation of crucial ecosystems.[4] Activists and NGOs have an important role in making private firms accountable for their actions related to environment. Even when there are regulations to mitigate pollution and other forms of degradation of environment, many private firms (especially in less-developed countries) may not follow them. It is in this context that the voice (agitations, protests and media campaigns) of NGOs matter. It forces the law enforcement officials to act, it may encourage governments to be alert and it may make the public at large to reckon these degradation issues. This may lead to a change in policies which have a bearing on environment (and land). We have mentioned the role of NGOs in carrying out innovative experiments in land restoration. Though one would expect public research organisations to be

[4] The case of Silent valley forest in the Kerala state of India is an example.

more active in innovation, the way resources are allocated and scientific research is carried out may favour a narrower set of solutions. Radical solutions may not be tried out initially by professional scientists. However, environmental NGOs may try out such solutions since they are not constrained by profits or the peer pressure among scientists. Though everything that they try cannot be scaled up, it may give on insights on feasibility, in large-scale land restoration. We have seen NGOs trying indigenous land cover for restoration without any concern for profits or sustainable income. Some of these NGOs may be concerned about the conservation strategies followed by the government (say the use of exotic species) and may try out alternative approaches.

There is an argument that the current lifestyle of people is not sustainable and hence there is a need to respect the limits of natural resources. Though there can be multiple views in this regard, society does not have enough information on the viability of alternative lifestyles. This is much more so since these alternatives should be sustaining/enhancing the utility or happiness of people. Otherwise, people may see these lifestyles as one depriving joy and hence their attractiveness may decline. Experimentation with alternative lifestyles is very costly, and may require active involvement or immersion of researchers in what is tried out, and hence, the information on this is less likely to come from normal researchers. However, a set of environmental activists are ready to try out these alternatives mainly due to their ideological and emotional orientations.[5] Though such orientations cannot be achieved easily in society, the experience of those who follow alternatives may provide useful information if there is adequate reflection on their life. This may give insights on what may be feasible in terms of alternative lifestyles without compromising the joy or happiness in life. This may be useful to think about long-run and sustainable solutions for land restoration.

Limitations/Challenges of NGO'S Actions

Most NGOs depends on money from others including international and national organisations/governments. Or they may depend on philanthropic organisations. There are difficulties for many NGOs to access these resources. Many times, informal connections are important in accessing funds but all organisations may not have such connections. It is obvious that the resources available with the NGOs are small compared to the global need for land restoration. Hence, it is unrealistic to see NGOs as a major player in land restoration but their importance lies in innovativeness and enhancing accountability of public and international organisations. However, the limited funding may work against the innovation potential of NGOs. They may become implementers of governmental projects, and in this case, they may be compelled to accept the narratives of governments in this regard.

---

[5] An example can be seen in https://navadarshanam.org/.

Most actions of environmental activists are of small scale in nature. This is so for different reasons. First, these experiments take place with limited resources mostly from public or altruist sources. There is also a tendency to limit the scale of environmental actions of NGOs due to a preference for higher controllability. Small is not only beautiful but are controllable. The probability of success in the way it is defined by the protagonist is higher in the case of these small experiments. However, these may limit the scalability of their actions. We have noted the need for large-scale restoration, and the usefulness of NGO's experiments and actions in this regard is limited.

One major problem with the actions of environmental activists is that they are driven by one or other ideology which may not be acceptable to common masses, or the rationale behind it may not be obvious to others. Or the success of their experiments can be seen only within the framework of this ideology. This may limit the replicability of their land restoration projects. For example, they may have specific ideas related to consumption (what may or may not be consumed) and production. We have seen conservationists who favour vegetarianism. When we conducted a workshop on land conservation, one group which promotes vegetarianism was an active participant. However, it is difficult to promote vegetarianism or expect to follow vegetarianism for land restoration. There are others who may consciously follow a less material lifestyle (including those who shun normal jobs in industry or services to take up restoration). However, such an approach is less likely to be acceptable to the majority. Hence, certain experiments of environmental activists may not provide replicable and feasible models of land restoration by considering the ideas and needs of the majority. This may work against the scaling ups of land restoration projects which are carried out by NGOs.

Another limitation of NGOs' actions to protect environment and land is with respect to the inadequate importance given to science and economics. We have noted in a chapter that the efforts to protect biodiversity in all projects of land restoration may increase the cost (including the opportunity cost) of restoration and that may work against the restoration of a sizeable area of land. There is also a need to be concerned about the real impacts of certain species like Eucalyptus on land. Ideological or emotional or aesthetic oppositions to restoration strategies are fine at the individual level but these may not be enabling for the design and implementation of large-scale restoration. There are individuals who may be willing to devote a major part of their resources and sacrifice material comforts for restoration, and that is laudable, but expecting the majority of people to follow these could be unrealistic. Large-scale restoration requires reasonable and realistic understanding of human behaviour.

Need to Be Complementary to Actions by Government and Private Sector

Based on an understanding of the pros and cons of NGO actions towards land restoration, one can argue that these have to be complementary to those by governments and private companies. There are certain disadvantages for the latter (both governments and private sector), and altruistic and innovative micro actions of NGOs have advantages in this regard. However, these cannot be scaled up, partly because of the fact that ideologies and altruisms of a set of individuals may not be followed by the rest of the society. The state has to consider the interests of all sections of society and some of these may not be adequately oriented to land restoration. Hence, large-scale restoration requires governments (and private companies) taking appropriate lessons from NGOs and use it for the design of their actions. NGOs can be watch guards of actions of governments and private companies. The criticisms of the former may encourage the latter to correct their actions. Hence, governments should not take an adversarial attitude, and instead take a constructive attitude towards the public actions of non-governmental organisations for this purpose.

Chapter 14
Role of Private Sector in Land Restoration

The interactions and discussions as part of the G20 GLI have demonstrated that land restoration is seen as a business by a set of private investors and companies. These companies have bought thousands of hectares of land and are converting these into plantations with indigenous species of trees. Their plan is to harvest the timber and other materials in a sustainable manner so that the land-cover remains intact, which is good for land restoration. Other firms buy raw materials which are collected by people from forests and indigenous plants and trees on a sustainable basis and make and market different products. This can increase the income of these people, and hence, it can enhance their incentive to protect these ecosystems. There are people who invest their money in creating or sustaining forests from which they expect a certain level of financial income over an adequately long period of time along with the protection of vegetative cover over the land. There are banks and financiers who facilitate nature-based solutions to address climate-change from which they expect income through carbon credits, and some of these solutions may strengthen land restoration too. There are possibilities and challenges in the entry of private firms in land restoration[1] and these are discussed in this chapter.

Benefits from the Entry of Private Firms for Land Restoration

Though the size of this land-restoration industry is small, these can bring in important benefits. Firstly, these firms facilitate the channelling of private capital towards land restoration. This is important when we note that the capital from individuals and that

[1] Some of these are noted in different documents including https://g20land.org/blog/the-essential-role-of-private-sector-corporates-in-advancing-natural-climate-solutions-and-ecological-land-restoration/; https://www.decadeonrestoration.org/publications/blueprints-private-investment-ecosystem-restoration-lessons-case-studies.

V. Santhakumar et al., *Land Restoration: A Rational, Grounded and Pragmatic Approach*, https://doi.org/10.1007/978-3-032-26652-1_14

of governments are not adequate to ensure the kind of land restoration that is necessary for the world. Moreover, if land restoration can attract profit-oriented capital, then it will flow almost automatically and regularly without much persuasion. Or if land restoration is seen as profitable, private companies will have the incentive to invest their capital for this purpose. On the other hand, the creation of such incentives by governments can be very difficult. Secondly, private companies tend to be more efficient since they operate with the objective of profit maximisation which builds on cost efficiency. There are challenges in achieving such efficiency (due to the lack adequate incentives) in government projects. The development of efficient practices by for-profit firms can benefit governments and other not-for-profit initiatives too. Thirdly, the entry of private sector into land restoration may facilitate innovation which can be used commercially or at large scale. Even when basic science and technology which are useful for land restoration are developed in public sector laboratories or universities, these may not be used commercially. On the other hand, the incentives for profit maximisation and cost minimisation may enable private or for-profit firms to invest in useful and commercially viable innovations in this regard. These innovations may not be limited to the processes of restoration. In addition, a wide variety of new products may be created for which the raw material may come from a restored land. Private sector efficiency in marketing may also enhance the consumer interest and market size of such new products. Hence, there are multiple benefits from the entry of private firms in land restoration.

Challenges

Since the business of these companies is land restoration, land is the most important input. However, the access to land is a major challenge in many countries. There are multiple ways to have access to land. Buying enough land is an option, but there may be possibilities to pool land owned by others (normal farmers) for the restoration business owned by private companies. There can be partnership with individuals or communities which own land. In certain cases, the private company may encourage land owners to produce products in ways that are in tune with land restoration, and these products may be processed and marketed by the former. However, there are challenges in all these routes to have to access to adequate land.

Buying Land—Difficult in Certain Countries

Buying land is very difficult for private companies. This is much more so in countries where land per capita is very small. The open market price may become very high and it may not be determined by the potential income from the use of land for agriculture and related activities. Instead, land may be seen as a speculative asset. The inadequate development of financial markets in developing countries may encourage people to see land as an investment good (for which the expectation is that there would be a secular increase in the price over time). This may encourage people not to sell their land even when they stop using it for agriculture. They may keep it fallow by expecting that its price would increase in future.

Given the low per-capita availability of land, the government may take steps to ensure that poorer people have access to land. This may be by bringing in restrictions against the transaction of land for non-agricultural purposes or towards commercial organisations. In certain cases, there may be land-ceiling regulations to see that it is not concentrated in the hands of a few players. Even when the economic conditions behind these regulations change, these regulations may persist and may cause difficulty in the buying of large stretches of land by private companies for restoration purposes. These regulations may affect the transactions and finally increase the effective cost to be paid for land.

In many poorer countries, private property rights are either not clearly defined or encounter weak enforcement. This can create an uncertainty in ownership and become another source of transaction costs. There can be other challenges too. In many countries, a sizeable part of the land is used by communities without clearly defined ownership for individuals. There may be restrictions on the transfer of land. In such cases, there may be a need for partnership with communities. But there can be coordination problems within the community, and this may work against reliable partnerships with outsiders. All these may increase the difficulties of private companies to have adequate access to land for restoration business.

Land is a resource that is contested in many parts of the world. The way future governments may behave in terms of the property rights of land is also uncertain. There could be attempts to restrict the rights of large private investors in many countries. This is much more so for international investors. These may discourage those want to invest in land restoration for financial returns.

Not Enough Finance

Since this is a relatively newer line of business, financial markets may not have developed adequate appetite for financing this business. The regulations which are made for financing conventional business may not be appropriate in this regard. For example, the public sector bank which has allocated money for land restoration in Brazil cannot fund most land-restoration firms since these are started recently. Hence, such firms may have to depend on own capital, and this itself can limit the scale of operations. Some of these firms see sustainable harvest of timber as a return from their investments, but then they may have to wait 50–100 years to recover the investments through this means. Any legal or regulatory uncertainty in the ownership of land may discourage investments of this kind.

Land restoration as an investment face other challenges, and these are similar what we have discussed in the chapter on finance in this book. Plantations of trees can give income only after 20–40 years. This requires adequately patient investors. There will be uncertainty in long-term economic trends globally and in specific countries. These uncertainties can discourage investors. Given the current interest rates in many part of the world which may influence discount rates, present value of future gains which may come after 20–40 years could be significantly less. Hence, there is a need to combine short-term and long-term gains. Carbon finance has the potential to give short-term returns of longer-term investments in land restoration but the institutional difficulties in this regard may pose challenges for the realisation of this potential.

There are enterprise-builders in land restoration but they need capital. The problems which we have discussed in the chapter on finance may reduce the flow of capital to these private enterprises.

Not Enough Trained Workers

Private firms which are involved in restoration business also face a scarcity of properly trained workers in the domain. A few companies conduct in-house training for this purpose. Conventional education programmes are for creating professionals who may be involved in land-degrading activities like intensive mono crop agriculture; Ecosystem restoration is yet to be an important part of the training in professional disciplines like agriculture or forestry. Even when scientists/practitioners of these disciplines get an exposure to land restoration, they may not be oriented to managing businesses to enhance the value of this economic activity. Conventional business education is not suitable to ecosystem (land) restoration as a business. Hence, the lack of sufficient human resources is a major challenge for these companies, and there have to be private and public, and global and national initiatives to address these challenges.

Technology a Challenge—Example of Large Scale Nurseries

There is a need to develop and use appropriate technology for enhancing the productivity of land restoration especially when it is seen as a private (for-profit) business activity. However, the development of technology in this space has not moved ahead compared to other areas such as commercial agriculture. For example, there is a need to have nurseries which can produce/supply seedlings of indigenous trees and plants in large scale. Though there are a few attempts in this regard by specific players (Regreen in Brazil is an interesting example), businesses in many other countries are yet to have access to such technologies. There is a need for investments in technology generation and also for programmes for technology-transfer for this purpose. Private firms in less-developed and poorer countries may not have access to these technologies and enhancing it could be a part of the agenda for international initiatives towards land restoration.

Scaling Up is an Issue Due to All These Challenges

There is a need to scale up the operations of these businesses. This is needed to achieve the global and national targets for land restoration. Moreover, such scaling up is needed to enhance the financial viability of each firm that is in the business of land restoration. There is a higher level of fixed costs (including that for professionals) in this business, and this requires enhancing the scale of operations. However, all the limitations which we have discussed here work against the scaling up of the operations of these firms. This is connected to procuring more land at competitive rates and accessing more capital. Hence, the current situation is that though such private business provide certain hope but its impact as a whole remains small.

We take up the case of a few private companies which see land restoration as a business in the remaining part of this chapter. This is based on field visits by Santhakumar during the last 2–3 years as part of his consultant assignment with the G20 Global Land Initiative.

Case I: The Case of Forest-Wise in Borneo Island

The main products of Forest-Wise[2] are made from the nuts of illipe tree (Shorea stenoptera). These trees are commonly seen in the wet tropical forests of Borneo Island. The butter of these nuts is processed to make a variety of cosmetic products (moisturiser, soap, massage oil, etc.) There are also other products of the company, such as Kukui nut oil, Arenga sugar, which are made from other materials collected from these forests. It works with about 1000 collectors of such forest products and its intervention could increase their income by 16 per cent. Its impact extends to 6 villages and 36,000 hectares of forests in the West Kalimantan area.

People in this area were using illipe nuts as an ingredient in some of their dishes. However, only a small part of the nuts available in forests adjacent to their village is used for this purpose. Forest-Wise could demonstrate that the butter of these nuts can be used for a variety of cosmetic and health products. It could also sell these products to domestic (in places, such as Bali) and international markets. The oil from these nuts can serve as a substitute for cocoa butter and palm oil, which are planted by destroying forests.

Hence, the intervention of Forest-wise encourages people in these communities to collect more of these nuts from adjoining forests. The illipe tree can be very tall, and its timber is also valuable. There are many such trees in forests here, and even if people cut grown ones for timber, many fruit/nut-bearing ones would still be available. The collection is done by picking fruits/nuts from the ground.

There is no scarcity of nuts in these forests. In fact, the main problem for the company is that they cannot buy nuts from many collectors. It is, as of now, buying these nuts from only a relatively small number of villages and even from these

[2] https://www.forestwise.earth/.

villages, it may not be able to procure the entire quantity that the locals can collect. The processing capacity of Forest-wise (and possibly, the demand for finished products which it makes) is limited. It is trying to get more investors so that the processing capacity can be enhanced. It may also require more concentrated efforts to expand markets to increase the number of customers for its products.

Though Forest-wise is a monopsony, it is giving a higher price for nuts to farmers/collectors by sharing a greater part of the surplus with the latter. The intentions to protect forests and share a greater part of the income with the locals are visible in the work of Forest-wise. The founders of the company do not see this as a profit-maximising enterprise, and forest conservation through sustainable means seems to be the main motivation. Though there were other companies sourcing illipe nuts in the past, before the entry of Forest-wise, these could not enter into reliable contracts with local people. There were unexpected fluctuations in the price. It is this problem that Forest-wise could address, though it could not source nuts from all people and villages where these are available.

Forest-wise is already involved in the production and marketing of different products, such as Arenga Sugar, Kukui Nut Oil and Bush Merah Oil in addition to those made from illipe nuts. The company is attracting other investors and hoping that their investments may help it to buy and process more nuts. This may help more people and villages. There is an interest in 'organic rubber' in European markets, and if this translates into sourcing of a sizeable quantity of rubber from multi crop/species plantations, it can reduce the mono-cropping of rubber trees in this region.

The life of a few villagers in the community also gives the impression that people there can have a decent life even without abandoning forests and cultural practices. Though there are companies like Forest-wise, these do not make much money. Some of the companies are still in an investment phase hoping for sustainable returns in future. Forest-wise is recovering the cost of its operations but may not have an investible surplus on its own to expand operations. Hence, they may not be in a position to scale up their operations. Hence, the operations of most of these firms currently depend on the goodwill of a few individuals who invest their money, time and probably life in this. The availability of finance to reduce CO2 emissions and protect biodiversity at the global scale has not benefitted these companies that much.

Case II: Private Investments for Land Restoration in Brazil

Private companies are in the business of land restoration in Brazil. We could visit two such firms namely Re.green (https://re.green/en/) and Symbiosis (https://symbiosis.com.br/). These firms and/or (wealthy) individuals are starting to buy land and using it for developing multi-species tree farms with plans for sustainable harvest. These restore lands even though such farms may not have all qualities of a natural or secondary forest. In addition to indigenous tree plantations, there are secondary forests also in these farms in Brazil based on the legal mandate to set apart a part of private land for restoration. There can be plantations of exotic species which also have secondary forests based on this legal mandate or due to a voluntary action to restore land. One such effort by BTG Pactual and Timberland Investment Group aims at creating eucalyptus plantations with about fifty per cent of the land devoted to natural regeneration. The vegetative cover over soil in these tree plantations can be left somewhat undisturbed, and hence, it can help the conservation of soil and water. The nutrients in soil are also enriched through this process. Hence, these plantations serve almost all purposes of land restoration other than the conservation of a higher degree of biodiversity. Moreover, such harvested woodlots can be as good as natural forests in terms of absorbing carbon dioxide. The fact that the price of agricultural land is relatively lesser (USD 3000–5000 per hectare in certain areas) in Brazil and land is available for agricultural purpose are enabling factors for these private tree farms.

There are multiple initiatives of this kind and some of these have plans for (or started) expanding the acreage. However, these firms are yet to make a positive financial return on their capital. This can be due to multiple reasons: Though such private restoration firms have acquired hundreds of hectares, the area may have to be much larger to get a return by considering the fixed and operational costs.

The harvesting of timber at a viable scale may require 20 or 30 years of plant-growth, and hence, it may take time. There is also a perception that illegal timber reaches Brazilian markets and that is dampening the price of timber. This can reduce the returns from legally cultivated and harvested timber. The cost of labour is higher considering other opportunities which are available in commercial agriculture or industrial/service sector in Brazil compared to other developing countries. The managers of these tree-farms mention the need to build up skills among workers but then the cost of retaining such skilled workers may be higher. These factors work against the scaling up of these tree farms. The income from non-timber forest products including eco-tourism is not that high for these firms, and hence, these have to depend solely on timber and carbon credit. However, the money from the latter is yet to become a viable source of income for these private businesses of land restoration.

Despite all these challenges, the enthusiasm of private companies to have indigenous tree plantations remains higher in Brazil. The returns from such multi-species indigenous tree plantations can be partly financial and partly social/environmental benefits. It is not unusual to see a set of wealthy individuals who invest their money for social/environmental benefits. The fact that this activity, though is common in developed countries like the USA, is catching up in Brazil and that many of these investors are domestic citizens, are interesting. This experience may provide important lessons to other emerging and less-developed economies.

Chapter 15
Role of Governments

What should governments do (and may not do) to facilitate land restoration needs a systematic thinking and elaboration. This is even more relevant since actions of all individuals and firms (along with the state) will have a bearing on the degradation and restoration of land. One can see different kinds of failures on the part of the state in this regard: For example, the state has a role to prevent negative externalities associated with actions of private individuals/firms and there can be such actions leading to land degradation. However, the state may not be adequately active in regulating or restraining these actions. There can be governmental actions which lead to land degradation. For instance, subsidies which may encourage landowners to degrade land are of this category. We discuss certain actions that the state should take to facilitate land restoration in this chapter.[1]

1. Regulate Private Actors to See That Land Degradation is Avoided

Actions of private individuals and firms may lead to land degradation, and the state's role is to prevent or discourage such actions. The extraction of materials in a manner that would cause excessive damage to vegetative land cover or top soil (and inadequate rehabilitation of the cover after this extracting activity) is an example. Ideally, the state should introduce regulations to avoid such actions. In most cases, one can see regulations in place but these are not effectively enforced. There can be multiple reasons for this ineffective enforcement. First, certain activities may be seen as important for economic development, and appropriate regulation of such activities may be seen as anti-development. Or private enterprises may view such regulation as anti-business. Secondly, since a large number of private individuals may take steps which cause land degradation, regulating their activities may require a huge enforcement costs. This happens when people themselves have the incentives to cause land degradation.

[1] There are advocacy notes on what governments should do for specific aspects of land restoration. Refer https://www.wri.org/insights/public-incentives-land-restoration-latin-america. This chapter takes an overall view based on the experience of our work as part of the G20 Global Land Initiative.

V. Santhakumar et al., *Land Restoration: A Rational, Grounded and Pragmatic Approach*, https://doi.org/10.1007/978-3-032-26652-1_15

Hence there are three steps which are necessary on the part of the state: (a) considering land restoration as an imperative and making appropriate regulations; this part is relatively easy and international organisations, including the G20 Land Initiative can be helpful in this regard; (b) synergising the objectives of economic development and land restoration; this is also possible if we understand the economic costs of land degradation, and the economic/financial benefits of pursuing a path of economic development that restores land; (c) ensure that people at large lose the incentive to use the land in a degrading manner. The latter is connected to human development which we will discuss in a following section.

2. Ensure That State Policies Do Not Lead to Land Degradation Indirectly

We have mentioned the case of public subsidies leading to land degradation in the past. These may include those for the excessive use of water (from canals or tube-wells) for cultivation, chemical fertilisers, pesticides, etc.

These policies are often driven by political economy (due to the influence of interest groups or the need to get votes from specific sections) without assessing the overall gains or longer-term impacts (even to those people who are supposed to be the beneficiaries). It is naive to expect that such political economy pressures will disappear. This is where public discussions on land degradation, the need for land restoration, its positive benefits for all including those who depend on land directly can make some difference.

In most cases, these subsidies can be redesigned without compromising the re-distributional objectives that these are supposed to serve. For example, farmers can be provided direct cash or income transfer instead of subsidised fertilisers or pesticides, which may encourage them to use these materials excessively. There can be an adequate investment for water management rather than excessive focus on large-scale canal systems which may cause desalinisation of land or on subsidies for the extraction of ground water.

Private firms can also be the beneficiaries of public policies which may lead to land degradation. However, the correction of such policies is relatively easier (compared to those which are expected to help a large number of people/farmers). There can be organisations and institutions which gain from the status quo and may have an interest in the persistence of inefficient policies. For example, fertiliser factories may gain from subsidies for chemical fertilisers. Identifying and changing these organisational/institutional incentives are also important.

3. Enable Human Development So That People's Dependence on Land Declines

The core lesson of this book is that inclusive economic and human development is necessary to restore land (since such a development is needed to reduce the dependence on land for a large section of society). Hence government actions which facilitate human development are important for its own sake but also for land restoration.

What are the steps which are needed in this regard? There should be an intervention to eradicate absolute food poverty. This is better attempted through a public distribution system that provides basic food materials at subsidised rates. On the other hand, the idea that all people should be supported so that they can cultivate their own food may not be a good. It may be better to achieve efficiency in food production, and enable a major share of people to move out of agriculture. This may require marketisation of food in general but a public distribution system for those who cannot afford to buy from markets (based on a procurement from those who produce the surplus).

It is important to ensure that all children have access to formal education. This is important for multiple reasons. Girls' education may reduce fertility rates and reduce the size of the population in the long-run. Education may encourage and enable a section of youth adult population to take up non-farm jobs even in rural areas (which, on average, have higher wages). This may enable a higher share of the population to depend on industrial and service sectors. (These sectors have a relatively higher rate of return compared to agriculture). Many may argue that when people get education in poorer countries, they may not get jobs within their localities. However, there can be opportunities for migration to cities or other countries. One can see an exodus of people even from poorer regions and countries in the current world. These may reduce the dependence on, and use of, land.

4. Finance Projects Which May Not Attract Private Capital

Projects that conserve land but also give a normal market return may be financed by private individuals or investors. The state should ensure that bureaucratic complexities do not dampen such investments. On the other hand, those projects which yield benefits to society and not much to those individuals who take action may require governmental intervention. However, most projects are somewhere in between. There may be certain gains to individuals who invest, but, in addition to that, there could be certain benefits to others in society. The latter is the positive externality. It is important for the government to understand this positive externality part.

Even if there is a positive externality, there are projects for which private benefits are greater than private costs. Individuals may have an incentive to invest in such projects, despite the positive externality in these cases. But the crucial cases are the following: Those for which social gains are higher than social costs, but private gains (for those who invest or take actions) are lesser than private costs. Individuals may not have an incentive to take up such projects. The restoration of land cover for carbon sequestration or the protection of biodiversity could be an example. State should prioritise investing in these projects/programmes.

However, there can be multiple failures in this context. First, governments may invest in those projects which may be carried out by private investors if they are allowed to so. State may not invest adequately in those cases where private actors do not have the incentive to do so. Many governments face the problem of lack of enough public resources, and in such a context, land restoration may become a low-priority expenditure. There has to be a change in this situation.

5. Facilitate Private Sector Investments Which Lead to Land Restoration

In our studies and field visits, we have seen a number of private sector initiatives to restore land. These may be driven by a combination of goals: profits, social recognition, concerns about society and nature or by the intrinsic motivation to do the right thing. However, as noted in a previous chapter, these initiatives face a number of challenges to scale up their operations.

The major constraint is the not so enabling policies of the government. In many contexts, there are restrictions against the buying and owning of land by private companies. Though some of these restrictions may be legitimate for distributional reasons in the past, these may be continuing even if these are irrelevant in the current context. Governments are not so pro-active in changing these restrictive regulations. There may be cases where governments restrict the sustainable use of products from restored lands. For example, private individuals and firms are not allowed to harvest timber of certain trees in India. This can disincentivise the cultivation of such trees.

In certain other cases, governments may announce financing mechanisms which can be tapped by private companies for restoration. But there may be procedural hurdles. For example, the rules of eligibility to get these funds may not be suitable for new and genuine investments for land restoration. We have seen such a situation in Brazil where the government and state-controlled bank have provisions to support private investments for restoration.

6. Ensure That Institutions/Laws and Their Enforcement Prevent Land Degradation and Enable Land Restoration

It is obvious that laws and regulations are needed to reduce negative externalities like land degradation. In reality most, including poorer, countries have these laws on paper. For example, there may be regulations on the discharge of pollutants in most countries. However, as noted earlier, the real problem is in the enforcement of these regulations.

In general, it is relatively easy to make regulations. In fact, there may be incentives for poorer countries to make regulation for the conservation of natural environment (including land). International collaborations and platforms which may provide these incentives operate in a carrot and stick fashion. Further, the enforcement of these regulations is left to the domestic political economy.

In many such countries, domestic constituencies which demand the enforcement of environmental regulations are relatively weaker and/or those who are affected by a proper enforcement could be stronger. For example, though the burning of peatland is a global concern and Indonesia is an active participant in international deliberations, the domestic constituency which may be affected by a complete ban on peatland burning is sizeable there. Domestic pressure may mediate the enforcement of environmental regulations in these countries.

Hence it is important to tighten the enforcement of environmental regulations. Advocacy for adopting new regulations may not be that effective in pushing these countries towards a higher level of conservation. There is a need to understand the interest of domestic constituencies in demanding enforcement. It may be possible to take a step-by-step approach and tweak incentives at the margin so that the goal of these environmental regulations are met.

7. Support Communities Wherever These Are Capable to Manage Land Sustainably

As noted in a previous chapter, there are communities which are currently managing common land and related natural resources (like forests) reasonably well (even though there are others whose performance is not so desirable). The mutual dependence of the members and the informal enforcement of community norms may be enabling the former. Though these communities see increasing demand for some of their products (and also experience a need to enhance their consumption), such a growth can be made compatible with the way they manage their land and forests sustainably. There is a need to identify these communities and ensure that newer regulations and governmental actions do not impede their own management of resources.

However, the government may have to step in where communities' capability to manage common resources declines due to one or other factors. When community norms break down, there is a temptation to replace these with formal laws and enforcement by governmental agencies. An alternative solution that is projected in such a situation is the privatisation of common properties. However, there can be undesirable distributional implications which can also affect their effectiveness. Moreover, the positive externalities associated with common property resources may be under-provided when these are privatised. Hence, the attempts to privatise CPR must be proceeded with careful scrutiny. However, an unreflective use of formal laws and government organisations may not improve the situation, or there can be inefficient outcomes. There may be a need to craft co-management mechanisms which can retain the strengths of community institutions but can mitigate their weaknesses which may arise due to social change.

8. Respond to and Learn from Actions of Non-Governmental Organisations

There are a number of initiatives by different non-governmental organisations for land restoration. These are mostly trying out innovative solutions in different locations. The resources for such initiatives are mobilised mostly from altruistic (or not-for-profit) sources. Most of these are driven by the love for natural environment or an ideological and/or concern. For this reason, it may be difficult to expect the large-scale replication of these initiatives. Most of these are small-scale experiments.

However, governments can take a facilitating approach to these initiatives. It should not create procedural or legal hurdles against such projects as long as these don't create negative externalities. Secondly, such initiatives may be able to come out with innovative solutions which can be replicated by governmental or private organisations in their effort to achieve large-scale land restoration. Non-governmental organisations which carry out innovative experiments may have acquired a higher level of expertise on various aspects of land restoration, and this can be used for capacity-building of different stakeholders for this purpose.

The other potential contribution of non-governmental organisations is the actions against the government whenever it fails in mitigating land degradation. Though these actions may be seen as adversarial, these can serve a useful purpose of compelling different government agencies to act. Moreover, the pro-environmental actions of NGOs can serve as a countervailing force against the lobbying and other influences of narrower but powerful interest groups which may not like the regulations for the restoration of land.

In certain cases, there can be non-governmental actions by the collectives of business organisations for land restoration. These may also help the spread of information on good practices among its members. Government can collaborate with these organisations too. There are examples of this kind in Brazil and India.

9. Facilitate the Collection of Appropriate Data on Land Degradation/Restoration

Land restoration practices and their monitoring require better quality data. However, the systems of data gathering and compiling for this purpose are weaker in many parts of the world. On the other hand, there is a potential to use newer technologies for this purpose. Combining geospatial (satellite) data collection with information technology (including artificial intelligence and machine learning) has the potential to make better quality data quickly available for this purpose. India has used satellite data to report the change in land degradation.

In certain cases, appropriate policies can generate data for this purpose. For example, the Brazilian law which mandates the conservation or restoration of a part of private land is expected to generate information on the status of land. Though there are gaps in this data and delays in the procedures to collect this data from all parts of the country, it is an interesting case where policies would encourage the generation of enough data.

10. Encourage Research to Bridge Knowledge Gaps in Land Restoration

There are gaps in research and knowledge-base in different disciplines which work against effective land restoration. Though there are developments in geospatial sciences and technology, and information technology (including artificial intelligence and machine learning), more adaptive research can be attempted to enhance their use in land restoration. There is a need to quickly and clearly identify areas of land degradation and monitor the efforts towards restoring them and make these information easily accessible to different stakeholders.

The potential impact of each species (both indigenous and exotic) on land (its degradation versus restoration) needs to be understood a lot more scientifically. Incorrect or ideological views seem prevalent in this regard. There may be a need for combining ecological and economic insights to arrive at efficient/optimal species diversity in multiple locations (rather than pushing one or two models or goals in each and every location). There can be possibilities of species selection wherein revenues from land and goals/targets of restoration can be maximised. There may be a need to identify less harmful exotic species which can be used in those places where the rejuvenation of indigenous varieties may be difficult. All these show the importance of a lot more inter- and multi-disciplinary research for land restoration. Governments have to play an important role in facilitating all these research.

11. Coordinate with International Actors to See That the Country Benefits from and Contribute to International Efforts

Climate change and the erosion of biodiversity have enhanced the global interest in land and ecosystem restoration. As noted in the beginning of this book, international organisations have started a number of initiatives in this regard. However, the response of national governments is somewhat mixed towards these initiatives. The countries of the developed world may not need international pressure or finance to take action towards land restoration in their own territory. However, governments of less-developed and poorer countries wherein land restoration may gain from international initiatives, may not use these initiatives optimally. There is scope for more active participation of these governments to make use of international initiatives to enhance land restoration in their territories.

Such pro-active participation does not mean the acceptance of mandates or pressure or resources without much reflection. Instead, the countries can articulate their interests strongly at the time of shaping these initiatives. The domestic priorities should reflect more on what is agreed upon in international platforms, and once these agreements are made, there can be effective efforts towards the realisation of targets. However, one may see somewhat passive steps in this regard. Declaration of optimistic agreements in international forums; inadequate domestic policies and projects; the acceptance of international resources; not so effective steps towards of the implementation of projects and policies. All these indicate that the pressure on national governments for land restoration should come within the country, and that may be connected to the economic and human development there.

Conclusion

The restoration of land is a positive externality that benefits not only the owner of the land but also the society as a whole. Hence governments have an important role in this regard. Governments all over the world may not be adequately oriented to the needs of appropriate land restoration; or there could be an under-provision of governance in this regard as part of the failures of governments in general. There is a need to change the status quo to accelerate the process of land restoration.

Land is mostly in the hands of millions of private individuals and firms. However, they may not have the full incentive to use land in a restoring manner. This is where the government has to step in. Hence, governance related to land restoration requires a higher level of sophistication since it should encourage a large number of private actors to take steps, and ensure that these steps are taken indeed. This usually happens only where the quality of enforcement of institutions is higher, and also when people at large have certain intrinsic motivation to restore the land. These may be lacking in many countries. Enhancing the quality of governance is important for land restoration. This may not happen quickly. Moreover, such a quality of governance can also be connected to domestic factors including human development.

Chapter 16
Role of Different Stakeholders: The Case of Peatlands in Indonesia

The burning of peatlands can add significant amounts of carbon dioxide to the atmosphere and can also lead to other problems, such as haze that affects airline traffic, health issues for the local population and a decline in working days. There is widespread concern about the frequent burning of forests and peatlands in Indonesia. The national government and global community have taken certain steps to address this problem.

There are a number of studies on the ecological features of peatlands,[1] environmental problems due to their burning[2] and also the effectiveness of governmental actions.[3] One article which looks at the institutional aspects argues that 'it is critical that community livelihoods are considered in the restoration effort, and it is imperative to ensure that communities have profitable livelihood options that are compatible with ecosystem restoration'.[4]

This chapter argues that though the restoration of peatlands has global benefits which outweigh the costs, the benefits accruing to the Indonesian population may be lesser than their costs. This may have implications: (a) there may be inadequate efforts within Indonesia to restore peatlands even if adequate laws are made for this purpose and (b) there is a need for effective international support.

---

[1] https://reliefweb.int/report/indonesia/why-peatlands-matter; https://www.unops.org/news-and-stories/stories/restoring-indonesian-peatlands-protecting-our-planet; For a review of the studies on the nature of peatlands in Indonesia, see Warren et al. (2017).

[2] https://gggi.org/wp-content/uploads/2020/11/Indonesia-Peat-Insight-Brief_October2020_tentative.pdf

[3] For example, Dohong et al. (2018).

[4] Yuwati, et al. (2021).

V. Santhakumar et al., *Land Restoration: A Rational, Grounded and Pragmatic Approach*, https://doi.org/10.1007/978-3-032-26652-1_16

Fire in Peatlands—More Severe Than Forest Fires

People who are not specialists may not have a clearer understanding of the nature of forest fires in Indonesia. Many may be thinking that ordinary forests are put on fire by people or companies and these fires cause haze and emission of carbon dioxide into the atmosphere. However, there is more to it. There is a need to have a deeper understanding of the socioeconomic determinants and implications of fire that happens over peatlands in Indonesia.

A major part of the land in certain provinces of Indonesia (which are known for frequent forest fires) is covered by peat deposits. Peat is a soil deposit with a higher content of dead wood (historical accumulation of wood materials from past vegetation). 'Peatlands are composed of thick layers of partly decomposed organic material that (are) formed over thousands of years, (and) they store lots of carbon'.[5] Since these soil deposits have a high carbon content, they release huge amounts of carbon dioxide when they burn. About 36 per cent of the world's peatlands are in Indonesia.[6] Nearly half of the total emissions of carbon dioxide from Indonesia (in 2015) are from peatlands. The depth of these deposits can be anywhere between 1 and 4 or more metres.

Whenever there is a fire, these peat deposits continue to burn even after the vegetation over the surface is burnt down. There can also be a burning of peatlands unconnected to forest fires. Depending on the depth of the peat deposit, the completion of its burning may take days and weeks. The severity of forest fires (in terms of the emission of carbon dioxide to the atmosphere for a significant time once a fire occurs) can also be due to the presence of peatlands. The peatlands are generally wet but drained peatlands are highly prone to fire. Extinguishing the fire on the surface will not lead to the cessation of the burning of peat soil. It may be noted that setting fire to vegetation can be a cheaper solution to make it suitable for cultivation in many contexts, but the presence of peatlands makes this a lot more environmentally harmful in these Indonesian provinces.

Peatlands—More Than Forests

In a province like Riau, 60 per cent of the land is covered by peatlands.[7] The total area of peatlands in Indonesia could be somewhere between 20 and 25 million hectares. Hence, peatland can be observed not just on forest land but also on agricultural,

[5] https://reliefweb.int/report/indonesia/why-peatlands-matter; https://www.unops.org/news-and-stories/stories/restoring-indonesian-peatlands-protecting-our-planet; For a review of the studies on the nature of peatlands in Indonesia, see Warren et al. (2017).

[6] Warren et al. (2017).

[7] Source: https://peatlands.org/assets/uploads/2019/06/ipc16p342-347a361karyantoismail.etal_.pdf.

industrial, residential and urban land.[8] The restoration of peatlands (or the efforts to avoid the burning of peatlands) has to take place in all these different kinds of lands or there can be fire even on soils which are not covered by forests. Hence, the problem may not be forest fires but land on fire. Even normal use of land for farming or industrial purposes has to be carried out in a way which should not cause the burning of its soil. The mitigation of this problem requires the coordination of different agencies; peatland conservation cannot be the responsibility of any one agency like the department of forests, which is the main agency for controlling forest fires.

Cost and Benefits of Preserving Versus Using Peatlands

Peat soil is highly acidic and not many trees and plants can grow in it. Pineapple and timber trees like acacia grow in peatlands but these may not be enough to give an adequate income to local people. This would mean that the benefits of conserving peatlands are not high for the local people. Moreover, the opportunity costs of conserving peatlands are higher since local people (especially in provinces like Riau) cannot use a major part of their land surface for other purposes (Kiely et al. 2021). The cultivation of normal crops requires clearing vegetation and the addition of anti-acidic materials. The easiest way to clear the peatland to make it suitable for cultivation is by setting it on fire.[9] Hence, the benefits to the local people or others in putting this land on fire are higher (even if it causes environmental and other problems). Otherwise, they may have to spend a lot more and also be cautious to avoid fires even on land that is not forest land. Given these costs and benefits, the local people may not have a high incentive to protect or conserve peatlands.

It is true that the burning of peatlands affects local people by causing local environmental problems. Local schools may have to be closed down when the fire is intense. There may be a loss of working days and an increase in illnesses. However, a major part of the environmental damage due to peatland fires is borne by other countries (like Malaysia and Singapore because of the haze it causes and its impact on activities such as air travel) and the world (due to the emission of carbon dioxide).[10]

It is globally beneficial to restore peatlands rather than incur losses due to their burning.[11] There are estimates such as the following which highlight the rationality of restoring peatlands: Kiely et al. (2021) note, '... the losses and damages under a scenario where 2.49 million hectares of degraded peatland had been restored, finding

[8] Source: https://www.unops.org/news-and-stories/stories/restoring-indonesian-peatlands-protecting-our-planet.

[9] Fire 'is perceived as the most cost and labour-efficient method for removing vegetation. It also temporarily improves soil fertility, reduces acidity and reduces pests'. Source: https://www.cifor.org/publications/pdf_files/infobrief/6449-infobrief.pdf.

[10] Source: https://www.vox.com/2015/10/30/9645448/indonesia-fires-peat-palm-oil.

[11] Kiely et al. (2021).

a reduction in economic losses of US$8.4 billion'. However, the issue is that local people may get only a small part of these benefits, and that may not encourage them to restore peatlands. If we take 1000–1500 USD per hectare as the cost of conservation,[12] the benefits in terms of reduced fires and CO2 emissions (and also benefits to the local population) may outweigh the cost, but that part of the benefits which is going to the local people alone may be less than the cost to them.[13] Even though the avoidance of fires is an important part of the benefits of peatland restoration (which according to one estimate is the avoidance of a loss of USD 19,000/ha, local people enjoy the direct benefits of fire (as a way of clearing forests for other activities). It may be noted that clearing forests without the use of fire would increase labour and equipment costs.[14]

The benefits of effective peatland restoration to the world as a whole outweigh the cost of restoration, and that justifies global support for ongoing peatland restoration efforts. Hence, peatland burning is an international negative externality. Though the fire on each hectare may lead to the emission of 50–70 tonnes of CO_2, most of this is borne by the global community as a whole (also in the long term) so, the local people may not have a great inclination to arrest this problem. Hence, peatland restoration requires global action.

The global community cannot blame Indonesian people for using peatlands in an environmentally unfriendly manner. In Indonesia and other contexts such as it, where the population pressure is high, many people depend on the land and natural resources for their survival. Even though there is a spread of education and some movement of people towards industrial jobs, most of the industries in the region also depend on land and natural resources.[15] The industries which are less dependent on land/natural resources (like in the service sector) are less developed in provinces which have major peatland areas.

[12] This is taken from https://gggi.org/wp-content/uploads/2020/11/Indonesia-Peat-Insight-Brief_October2020_tentative.pdf.

[13] To some extent, this is acknowledged: 'Economically viable and environmentally sustainable livelihood options for smallholders in Indonesian peatlands are limited, underdeveloped and urgently need to be enabled and expanded.' https://www.cifor.org/publications/pdf_files/infobrief/6449-infobrief.pdf.

[14] Guyon, A. & Simorangkir, D. The Economics of Fire Use in Agriculture and Forestry—A preliminary Review for Indonesia. in *Project FireFight South East Asia* (Project FireFight South East Asia, 2002).

[15] This seems a problem of the industrialisation of Indonesia as noted by other commentaries. For example, see https://unctad.org/system/files/official-document/BRI-Project_RP07_en.pdf.

Table 16.1 Peatlands under different control/ownership and strategies for restoration

Land type	Means of restoration	Agency	Challenges
Conservation areas	Banning use	National ministry of environment and forests	Limited resources of the ministry
Forest for productive purposes	Regulation	Provincial departments	Limited resources; development needs/ compulsions
Concession area	Regulation	Ministries which deal with private investments	Enforcement challenges
Land used by communities and people	BRGM	Direct support	Limited resources

Different Kinds of Lands in Indonesia (and Responsibilities for Restoration)

We can categorise lands which have peat deposits in Indonesia into four types. The first one is conservation areas (forests), which are directly controlled by the National Ministry of Environment and Forests. Then there are forests for productive purposes which can be under the provincial department of environment and forests. There is the third type, which can be used for productive purposes and is leased out to private companies (there are also lands of this type which are not leased out). Finally, there are lands, which are operated by communities and people.

The conservation of peatlands is expected to take place through different means in these different types of lands (Table 16.1). Though Peatland Restoration Agency of Indonesia (BRGM) has a role in all these lands (say, for example, by providing financial support to the agency managing national parks and other conservation areas), its direct role is somewhat limited to those which are not controlled by other ministries and private companies. However, it is the only agency which is responsible for peatland restoration in areas used by communities and people.

Efforts Made by Indonesia

Indonesia is concerned about the negative impacts of forest fires on its own people and also the world as a whole. It has taken several steps to mitigate the problem. A notable part of forests in Indonesia is declared as conservation areas and it is under the management of the National Ministry of Environment and Forests. There is a notable reduction in deforestation in Indonesia.[16] No productive activities are allowed in

[16] 'While absolute forest loss remains high (340,000 hectares/840,000 acres in 2018), the direction of travel appears unmistakable. Indonesia's success is in part due to the robust measures President Jokowi's government has put in place, including temporary bans on further expansion of oil

these conservation areas. The government has instituted a moratorium on leasing out peatland areas to private companies in 2011, which was renewed afterwards.[17] There has been an increase in vigilance and patrolling in coordination with local communities so that there is a quicker collection of information on and actions against forest fires. Similarly, there are provincial governments which also control and regulate the use of forests. There are regulations on what private companies can do when they lease and use land in Indonesia for different purposes. Through the efforts of the private sector, which is monitored by the ministry of environment and forests, it is estimated that 3.64 million hectares (9 million acres) of peatlands have been restored by the end of 2020.[18]

Extinguishing fire as soon as possible requires a coordinated approach and cannot be accomplished by any one agency. Hence, there is a task force at the provincial level incorporating military, police and functionaries from various departments to respond to forest fires. There is also a coordination mechanism at the provincial level. It seems that the frequency of forest fires has declined in Indonesia during the last few years, though it is not known whether there could be an increase in fires due to weather-related events like El Niño.[19]

However, the persistent occurrence of forest fires, mainly in the peatlands has encouraged the Government of Indonesia to take more focussed actions. One such action is the formation of the Peatland Restoration Agency (or BRG initially, which has become BRGM later) in 2016 with the task of restoring 2.6 million hectares of degraded peatland (which included 1.7 million hectares inside concession areas and the rest outside it).[20] The establishment of this agency itself reflects the political commitment at the national level to restore peatlands. This agency is expected to mobilise resources from international and national sources, coordinate with different governmental and non-governmental entities, and facilitate peatland restoration in different types of lands. In 2020, it got an extended mandate to restore 1.5 million acres of degraded mangroves too.

Strategies of the Peatland Restoration Agency (BRGM)

Peatland deposits remained as they historically had since they have a high level of moisture. However, this moisture level comes down as part of human activities, for example, when a canal is made (the water in the soil deposit drains into the canal).

palm plantations into forests and peatlands.' https://www.wri.org/insights/5-step-plan-protect-and-restore-indonesias-forests.

[17] https://www.cifor.org/publications/pdf_files/infobrief/6449-infobrief.pdf.

[18] https://news.mongabay.com/2021/01/indonesia-renews-peatland-mangrove-restoration-agency-brgm/.

[19] https://education.nationalgeographic.org/resource/el-nino/.

[20] https://news.mongabay.com/2021/01/indonesia-renews-peatland-mangrove-restoration-agency-brgm/.

This dryness of peatland can lead to its burning even if there is no forest over it. Factors such as these make the response to the burning of peatlands a lot more complex.

The BRGM essentially follows a three-pronged approach to the restoration of peatlands. The first is rewetting. There are areas where peatlands get drier during non-rainy seasons, and this is aggravated due to man-made actions, like the digging of canals. Hence, BRGM funds different agencies, including local communities, to create check-dams across these canals so that the water level goes up to keep the peatland wet.[21]

In addition, it funds the digging of deep wells so that there is enough water to be used to keep peatlands wet when there is a fire. The second strategy is revegetation. Since peatlands sustain only a few types of trees and plants, BRGM provides seedlings to local communities to enhance the vegetative cover over degraded peatlands. It also promotes pineapple cultivation as an intercrop for a certain duration as a way of ensuring some income for local people who help with the restoration of peatlands. The third strategy is revitalisation, which is aimed at enabling the livelihoods of local people in ways which do not lead to the destruction of peatlands. The cultivation of pineapple and its marketing to distant markets are part of this strategy. In addition, BRGM also supports local people to establish aquaculture units so that their dependence on peatlands decreases. We looked at projects of all these kinds. These are useful and can make a positive impact. One account states that '... by the end of 2020, the BRG had managed to restore 835,288 hectares (2.06 million acres) of peatland outside concession areas or 94% of its target.[22] However, its impact in concession areas is limited.

Persisting Challenges

Though different agencies of the Indonesian government including BRGM take actions to restore peatlands, these are not adequate. The BRGM has the resources to invest only in a part of the projects which are needed to keep the peatlands wet and to have replantation or revegetation over degraded land. In one location that we visited, the officials of BRGM constructed one canal block (check-dam) where, according to them, 4 or 5 such structures are needed. The cost of such check-dam is not low. Similarly, the support for livelihood activities is given only to a smaller share of local communities which may potentially use peatlands in a degrading manner. If BRGM has to invest money for all such purposes, it may require substantial additional resources.

---

[21] This is a solution tried out by other agencies too. For example, USAID and UNEP attempted such a strategy in the central Kalimantan part of Indonesia. https://www.unep.org/news-and-stories/story/unep-supports-project-restore-peatlands-indonesia.

[22] https://news.mongabay.com/2021/01/indonesia-renews-peatland-mangrove-restoration-agency-brgm/.

Inadequate resources create other challenges too. Though a significant part of the land is recognised as national conservation areas (like national parks) to be managed by the National Ministry of Environment and Forests, and there is an increase in the presence of its personnel on the ground, this is not adequate to prevent illegal encroachment or other activities which degrade forests/peatlands. Forests which are used for productive purposes, and which come under the provincial department of environment and forests also face a similar issue. These organisations in provinces like the Riau get financial support from BRGM for peatland restoration, but these resources are also not adequate. Though there are regulations on what private companies can do in forests/peatlands, the efforts to monitor their activities are not adequate. Hence, they continue those interventions which are not suitable for peatlands. When blocking canals is recognised as a strategy to restore peatlands, there is a need to ensure that no more canals are dug for this purpose. This is also connected to the challenges of enforcing regulations.

The challenges are not merely due to inadequate resources. Since peatlands cover a major part of the fire hotspots, and the continued dependence of people on land (for small-scale agriculture or land-based industrial production, including plantations), the domestic political economy may work against the restrictions on the use of these lands. It will be difficult to have a complete ban on the use of peatland in such contexts. Even if there is such a ban, there will be political-economic pressures to allow certain activities, legally or illegally, even though some of these activities can lead to forest/ land fires. Moreover, such a political economy will make the cost of enforcement of regulations very high. Though BRGM is established with a greater focus on peatland restoration and under the direct control of the President of Indonesia, its authority and effectiveness over areas which are controlled by private companies seem limited.[23] The development needs may also force the Government of Indonesia to make regulations which may work against the restoration of peatlands. It is noted that the programme to increase food production encourages the government and the ministry of environment and forests to adopt regulations which may lead to the clearing of protected forests.

The need for economic recovery may also encourage the government to adopt laws which may facilitate timber extraction and other activities.[24] Though environmentalists and activists may criticise the government of Indonesia for these steps, we can interpret these as reflections of the unavoidable developmental needs of the country. Moreover, the assumption that the adoption of conservation-oriented laws by the government would lead to the restoration of peatland is unrealistic due to the challenges in enforcing these laws, which again is a reflection of the problems of development that Indonesia faces. There are arguments for meaningful policies

---

[23] This is noted by a few studies. For example, https://iopscience.iop.org/article/10.1088/1755-1315/995/1/012068/pdf.

[24] Apparently, the omnibus law 'clearly facilitates timber extraction and forest exploitation in the name of national economic recovery, citing a provision that scraps the requirement for all regions to maintain a minimum 30% of their watershed and/or island area as forest area'. https://news.mongabay.com/2021/01/indonesia-renews-peatland-mangrove-restoration-agency-brgm/.

and governance reforms for successful peatland restoration.[25] However, there is not enough reflection on the possible connection between development and governance and enforcement failures with respect to it.

Need for Public Investments for Peatland Restoration

Technical solutions for the mitigation of peatland fire require substantial investments, and these may yield benefits only over a long period of time. The protection of peatlands may require avoidance of drainage for agriculture, commercial forestry, peat extraction and infrastructure development in these areas. Moreover, there may be a need to mitigate the effects of global warming to see that it is not leading to the drying of peatlands. It may require ensuring that the soil deposits remain wet as a whole throughout the year.[26]

Hence, there is a need for substantial investments if a governmental agency decides to conserve the remaining or the most vulnerable peatlands of the country. The national investments that are required for this purpose are huge. Like any other developing country, Indonesia faces multiple demands on its limited public resources and hence, it may not be able to allocate enough resources for peatland restoration.

The Political Economy Problem of Peatland Conservation in Indonesia

Unlike other global environmental problems, the peatland fire poses certain additional challenges. There are environmental losses due to peatland fires but a major part of these is borne by other countries and the world as a whole. Though local people also face losses due to it, these may be relatively smaller than the loss to the world as a whole. On the other hand, the direct and indirect costs of the conservation of peatlands are much higher for the local people, as discussed earlier. Even if the government of Indonesia is compelled to take action, it may face many challenges due to this situation (and also the higher opportunity cost of public resources). The per-capita income of most Indonesians may not enable them to have a higher willingness to pay for the restoration of peatlands. It may be noted that the development status of Indonesia may not enable its people to voluntarily contribute (make the effort) to restore peatlands for the benefit of the world when such restoration does not give them adequate benefits (or when the opportunity cost of restoration is high).

The severity of this problem may become clearer when we compare it with other such ecosystems, like rainforests or mangroves. Rainforests may give a higher

---

[25] Dohong et al. (2018).

[26] These technical solutions are discussed in detail in this report. https://luk.staff.ugm.ac.id/rawa/GiesenNirmala2018TropicalPeatlandRestorationReportIndonesiaForBRG.pdf.

income to local people if they are managed sustainably. Here, the challenge is to encourage people to move towards sustainable management. (However, such sustainable management of peatlands may not be that beneficial to local people). The fires in mangroves may not be that severe since by definition, these are located in wetlands. For all these reasons, the challenges to conserving peatlands could be more severe.

In a context where the cost and benefits to the local economy are tilted towards the degradation of peatlands, one approach is to argue for a greater focus on the sustainable livelihoods of local people as a strategy to restore peatlands.[27] Though opportunities of this kind should be explored, our impression is that the possibilities in this regard are limited. Moreover, the need to have higher incomes may encourage people to depend on land/forest-based industries in the region since other industries which are less dependent on these natural resources are less developed. In our view, there is a need for adequate international support for the restoration of peatlands.

---

[27] https://forestsnews.cifor.org/70022/peatland-restoration-in-indonesia-why-do-people-matter?fnl=en.

Part IV
Laying the Ground for Land Restoration—The Way Forward

Chapter 17
Need Education Programmes for Land Restoration

Education can play a certain role in changing the behaviour of human beings. First, it can enhance the awareness of people on the possible implications of their actions. Secondly, it may enhance their information on different opportunities and some of these can reduce the negative implications of their actions without a significant compromise in terms of their goals. Though the impact of environmental education on sustainable development is not clear, as evident from a number of studies, there can be certain changes which may facilitate such development if these do not lead to a substantial reduction in the satisfaction/happiness of people.[1]

Though other aspects of environment including those related to climate change are part of education at different levels, issues of land degradation or restoration are not adequately reflected in education. On the other hand, land is taken into account in both general and professional education (say in agricultural sciences) as an input or means to economic development. Here, the focus is on the utilisation of land to enhance the benefits for human beings. Such an approach may often overlook the negative impacts of certain ways of using land, especially when these impacts are to be felt by future generations. Hence, there is a need to include lessons from land degradation and possible ways to use land in a sustainable way at all levels both in school and higher education, and also in general and professional education. Specific details of such an education are touched upon in the following paragraphs.

[1] For a discussion on the possible impact of education on sustainable development, refer https://practiceconnect.azimpremjiuniversity.edu.in/do-higher-levels-of-education-make-people-environment-friendly/.

V. Santhakumar et al., *Land Restoration: A Rational, Grounded and Pragmatic Approach*, https://doi.org/10.1007/978-3-032-26652-1_17

Land Restoration in School/General Education

Land is discussed often in general school education. Its importance to the life of human and other living beings, and also in social, economic and cultural aspects are touched upon in different subjects in school education. As part of modernisation and economic development, there was a tendency to use education to encourage new generation to acquire 'scientific' ways of using land which may include the enhanced use of irrigation, high-yielding varieties, fertilisers, pesticides, etc., to increase the yield from agriculture. Such messages can be seen in school text books too. However, there is need to highlight that land is not an abundant resource. Its quantity can also come down say as part of sea-level rise due to climate change. But more importantly, its quality can come down drastically reducing its ability to provide a variety of goods and services and other attributes to human beings. An audit of the contents of school education to evaluate how land is discussed may be an important step. Additionally, this must be complemented with ways to ensure messages about its sustainable use. This is relevant in all parts of school education and not only in specific subjects such as environmental education.[2]

Land Restoration in Professional Education

Agricultural Sciences

Agricultural sciences have traditionally focussed on crop productivity. That was the reason for a higher level of research on high-yielding varieties, and the use of chemical fertilisers and pesticides. Such intensive agriculture is likely to flourish in irrigated fields, and hence, there is a focus on irrigation too. All these can enhance land degradation. Though alternative approaches, such as agro-ecology, organic farming, are popular these days (mostly among environmental activists outside public and private research stations), these are not integrated adequately into the formal research system. Moreover, given the practice of these as alternatives, the 'scientific and economic' aspects of these are underdeveloped. For example, there are several unanswered questions: What may happen when organic farming takes place in a major part of the land? Will there be an increase in prices due to any possible reduction in crop-production or increase in cost of production? Since all types of lands may require land restoration, and farming may have to be continued in many parts of the world, what can be the strategies to ensure land restoration without sacrificing crop production? Though there could be knowledge on these aspects, it is yet to become part of the education of agricultural professionals. Such an education is important for land restoration.

[2] G20 Global Land Initiative advocates such an approach. Refer https://g20land.org/event/nurturing-rootsthe-importance-ofintegrating-landrestoration-in-schools/.

Forestry

Conventional forestry has changed from its focus on timber extraction to ecological conservation. However, certain aspects of its practice is yet to become suitable for land restoration. First, the conservation of biodiversity or primary or secondary forests may be feasible only in part of the land of a country. The other part of land may have to be used for economic activities without causing land degradation. Forestry science can contribute towards such a land-use practice. For example, one or more indigenous and exotic species may help land restoration and carbon sequestration, even though it may not help the conservation of biodiversity. Though the creation of such plantations by cutting down natural forests may lead to net losses in terms of ecological services, there could be a gain when such plantations come up in degraded lands. Moreover, there may be cases where non-tree plant or biological growth is the appropriate way of restoring certain kinds of land. The conservation and management of such lands are yet to become part of forestry science. Hence, land restoration has to be an important focus of the education and research in forestry science.

Civil and Environmental Engineering

Civil engineers are involved in building construction and land and water management. Traditionally, the focus of these professionals was on the technical efficiency of man-made structures. They may be concerned about the optimisation of financial resources for these man-made activities. However, the need to avoid (or cause minimum) degradation of land has to be an important concern of these engineers. Even when land is to be used for construction and other activities, there may be possibilities of rehabilitation after its use. All these strategies are needed for land restoration, especially in a context where human settlements have become an important driver of land degradation.

Architecture, Urban Planning, Landscaping

Architecture may have to orient more towards green solutions. This may require the optimal use of land, water and energy for a given built space. Tree/plant cultivation in urban areas has not been a major focus of urban planning sciences. There are interesting developments in this regard. Miyavaki forests is one among them.[3] It can help the creation of native forests within a relatively shorter period in relatively small stretches of land. In this case, seedlings are planted at very high densities. This replicates the regeneration process that occurs in a natural forest when the canopy opens up due to a larger tree falling. The saplings grow very fast to compete for the

[3] https://www.creatingtomorrowsforests.co.uk/blog/the-miyawaki-method-for-creating-forests.

light and then natural selection will favour the fastest growing individuals and act to thin out the trees. This can lead to a dense forest in 20–30 years as against 150–200 years through natural process. There is also a recent focus on urban agriculture (Bailkey and Nasr, 2000). It is 'an industry that produces, processes, and markets food, fuel and other outputs, largely in response to the daily demand of consumers within a town, city or metropolis, in many types of privately and publicly held land and water bodies (which are) found throughout intra-urban and peri-urban areas. Typically urban agriculture applies intensive production methods, frequently using and reusing natural resources and urban wastes, to yield a diverse array of land-, water- and air-based fauna and flora contributing to food security, health, livelihood and environment of the individual, household and community' (Smit et al., 1996). These include small gardens, community gardens, aquaponic farms, inside-gardens, rooftop gardens, vertical farms, etc. These may enhance the availability safe food within cities but can also enhance the carbon dioxide absorption. Moreover, it can enhance the quality of urban environment.

Similar techniques can be used for urban forestry. The need for open spaces and urban parks are increasingly felt all over the world. There is a need for technological and managerial changes in this regard. This may require professionals to have a greater orientation to these possibilities. For all these reasons, architecture and urban planning may have to orient more to the needs of land and ecosystem restoration.

Training for Specific Category of Workers

Workers for Nurseries

One major chalenge that large-scale restoration faces globally is the need to generate millions of seedlings of indigenous plants (including trees). Dependence on small-scale nurseries is not only ineffective but also can increase the cost of operations in large projects. However, there is a scarcity of trained workers in such large-scale nurseries. Since these nurseries may have to use mechanised production and standardised procedure, their workers have to be skilled (and the current practice of using less skilled ones for this purpose may not be enough). This may require large training programmes. Or nursery worker/technician may have to part of formal vocational training programmes which happen after the basic education. There is expertise in this regard in specific organisations like the Re-Green in Brazil, and there have to be international efforts to use their experience, especially in other less-developed countries.

Education for Practitioners

Ground level-action in land restoration requires a large number of practitioners beyond individuals like farmers who may try to restore their own land. This is needed to facilitate large-scale action. This critically depends on up-skilling of these practitioners. For example, those who facilitate land restoration should be able to assess its impact on outcomes such as carbon sequestration, and be able to attract the financial resources available for this purpose. We understand that such expertise is not adequately available especially in less-developed countries. This may require capacity-building programmes for all kinds of practitioners—of both governmental and non-governmental organisations. We discuss some specific issues with respects to these two types of organisations in the following sub-sections.

Non-governmental Organisations

Most people who start or work with non-governmental organisations are highly committed to the restoration/conservation of land. Hence, there is a higher level of motivation. However, that is not enough to be effective in achieving land restoration to the level that is necessary by considering the global challenges in this regard. Large-scale land restoration requires mobilisation of adequate capital and human resources, and putting them to effective use so that certain desirable and tangible outcomes can be achieved within a specific duration. These require a lot more professional and managerial expertise in addition to intrinsic motivation and commitment. Secondly, land restoration is not something that is to be achieved in certain parcels of land (say forests). It is to be achieved over the whole land surface, including that is used for different human purposes. This may require a design of different strategies by considering the trade-offs between different human needs, and that of land restoration. This also require a critical reflection of ideas and thought processes of NGO activists who are involved in land restoration. Both these aspects have to be part of an education or capacity-building programmes of NGO functionaries.

Government Officials

Though what we have described in the previous section is relevant for government officials who are implementing land restoration projects, they may face other challenges too. They are more likely to be involved in discussions with international organisations and in the design of projects which intend to use global finance for mitigating climate change or conserving biodiversity. Hence they should have the technical capabilities to assess the marginal contribution of land restoration projects to carbon sequestration or biodiversity, and furnish details to financial agencies. Moreover, they should be in a position to design projects which ensure land restoration and also contribute to other specific economic and/or ecological needs. Government officials who are involved in such projects in less-developed countries may come from line departments of agriculture or forestry and they are not adequately oriented to the science and planning of land restoration. There have to be changes in this regard to plan and execute large-scale restoration projects.

Education for Policy Makers

Senior Government Officials

Most issues which are discussed in this book are relevant for senior government officials who are making policies related to land restoration. However, there is no specific set of officials in most governments who are entrusted with the task of land restoration. They may be working in agriculture, forestry or land revenue departments. Though there are officials who design economic and development policies, they are less likely to have an exposure to issues of land degradation. The lack of an overall view may limit the policies and actions towards this purpose. Though it may be difficult to create professionals or government functionaries solely for land restoration, its importance and the inter-relatedness of different factors that may enable/disable it, should be known to the majority of government officials, especially those who work in agriculture, forestry, environment, land administration, industrial promotion, social sector, economic development, planning, etc. These can be part of the administrative training programmes of government officials in different countries.

Officials of International Organisations

International organisations are involved in creating awareness and also providing supplementary funding for land restoration projects. However, these projects can be less effective, if the country-specific socioeconomic factors are neglected. The officials of these organisations have to be aware of the inter-connectedness between land degradation and socioeconomic conditions, especially that of less-developed countries. This book is an attempt to develop knowledge in that direction. The key ideas which are discussed here may have to be part of the capacity-building programmes of functionaries of international organisations which are involved in promoting land restoration projects in different parts of the world.

Education for Managers in Private Companies

As noted earlier, private companies contribute to land degradation. This can be severe in the case of activities such as mining or large-scale plantations. However, even those which may not use land as a major input into their production may contribute to land degradation directly or indirectly. In certain cases, there may be specific professionals or departments which deal with issues of land. But their actions or proposals need not get the full support from other managers. For all these reasons, managers of private companies need an orientation to land restoration. These may include the following aspects.

1. On the importance of avoiding land degradation: Many managers may not know that some of their actions can be harmful to the land. Hence, a general awareness on the importance of land restoration is to be part of education of mangers.
2. On regulations to minimise land degradation: There are regulations to avoid land degradation, and this may be known to professionals of that domain. However, an awareness of such regulations is important for all managers.
3. Good practices by private firms: Even for activities such as mining, there are good practices of rehabilitating land after its use. This may not be known to many organisations in poorer and less-developed countries. The managers of these organisations have to get an exposure to international good practices.

Education for Managers of Private Firms Which See Land Restoration as a Business

We have noted earlier that a set of private companies emerge and operate in the field of land restoration. They may view this sector with the potential to generate a combination of profits and social and environmental impacts. However, there is a need for education and capacity-building of the professionals of these companies, especially in three specific areas. First, is in the mobilisation of capital for their business. There may be a need to tap international and national capital that are aimed at addressing climate change and conserving biodiversity for this purpose. Using carbon credit mechanism may involve complex assessments and documentation. This may require expertise which is beyond that exists in most managers of such private companies. Second area is in the interaction with communities. This is relevant since land and forests in many parts of the world are used by communities, and there may be a need for partnerships to develop restoration-oriented businesses in these areas. Private company managers may not have the experience and capabilities to deal with communities and arrive at mutually beneficial relationships or partnership. Thirdly, private-sector activities in land restoration can have certain environmental impacts which demand innovation to reduce these impacts. Managers of these companies may have to be sensitive to such impacts. All these may require capacity building of managers and other professionals of firms which see land restoration as a business. This can be taken up by business schools and others which focus on natural resource management and private-sector operations.

Education in Specific Disciplines/Themes

Financing for Restoration

There are challenges in using profit-oriented financial instruments for restoration. Some of these may persist even when altruistic or social impact considerations exist for this purpose. There is a need for new instruments and strategies for this purpose which take into account the specificities of the land restoration. However, this may require education or training for finance professionals. This is more suitable for the new cadre of finance professionals who participate in what is called impact investing. Some of them may require an orientation to the possibilities and limitations of land restoration as an investment opportunity. These can be part of executive education programmes in management schools (especially for finance professionals).

Environmental Economics

Economics of land restoration has not received adequate attention among environmental economists. Environmental economics as a discipline has focussed more on pollution with land degradation being overlooked (though aspects of pollution may contribute to land degradation). This could be partly because the normal land (which is used for one or other human needs) may not be seen as an ecosystem worth protection through environmental regulations. However, there has to be a greater focus on land degradation and restoration in the discussions on environmental economics.

The conventional focus of environmental economics is the valuation of environmental losses, and also the design of instruments to reduce pollution or environmental damage. Though the estimation of losses is important, and it is used in the case of assessing the benefits of land restoration, there is not enough attention on the design of instruments which can lead to land restoration. Conventional instruments such as taxes or standard or quotas and so on may not be that relevant in the case of restoring land. Moreover, there is a need to take note of the developments in behavioural economics while thinking about the incentives or disincentives to restore land. There are not many efforts in this regard. This book makes a modest attempt in this direction, but there is a need to bring in such issues into the discussions and education of environmental economics.

Though there is a certain understanding of the possible connection between economic (and human) development and environmental pollution/conservation in environmental economics, this connection is a lot more important in the case of land. When societies are less developed, their dependence on land for basic subsistence is higher. As part of economic development, the share of population which depends on farming may come down. In poorer countries, people may depend on small-scale farms. The implications of all these on land degradation and restoration need to be thought out theoretically and also researched with appropriate empirical evidence.

The trade-offs between land restoration and other environmental goals such as the conservation of biodiversity and absorption of carbon dioxide need to be analysed carefully. This is because of the importance of land for a variety of human needs. Though land restoration can contribute to the conservation of biodiversity in certain contexts, there may be a need to restore land even if that may not enhance biodiversity, as discussed in a previous chapter. There may be a need for interdisciplinary knowledge of agriculture/forestry and economics for analysing these trade-offs. Certain ideological approaches (such as the one which give prominence to ecology) may not be enough to think about appropriate strategies for land restoration. For all these reasons, there have to be a greater focus on land restoration in environmental economics.

Ecology of Land Restoration

Land restoration has to be a focus of ecology as a discipline. Usually, ecology focussed on ecosystems which have had lesser interventions by human beings. Hence, natural forests, marine environment and mangroves have received greater attention of ecologists. Systems wherein human beings and their actions are an important part are yet to get adequate attention of ecologists. Land is such a system. Historically, land has been used for a variety of human needs. Its use for human needs has not come down in most parts of the world. It is almost impossible to preserve land in its pristine form whereas this may be possible in the case of remaining forests. How can there be an improvement in land as an ecological resource by continuing to draw other benefits from it is a challenge that is yet to get adequate attention on the part of ecologists. That is the reason why ecologists tend to consider almost all goals of restoration other than the conservation of biodiversity (which may require the cessation of almost all agricultural or industrial activity in a stretch of land) as somewhat inferior. It has prevented them from coming out with standards and strategies to restore land which are currently used (which will be used) for agricultural, industrial and urban needs. These are left to agriculturists or industrial and urban planners. For all these reasons, there has to be a reorientation of ecology towards land restoration.

Chapter 18
Conclusion: Connecting Dots on Land Restoration

There is a Momentum Towards Land Restoration But Needs a Lot More Grounded Actions

Different countries and international organisations are alert to the problems of widespread land degradation and the need for conscious efforts towards the restoration of this important resource. There are a number of global initiatives for this purpose, and the G20 Global Land Initiative is the latest one. Many countries have adopted targets and commitments in this regard. In addition, non-governmental actors also take notable efforts and innovative projects for land restoration. Communities in different parts of the world also attempt to ensure that the land—the crucial resource that sustains them—is conserved. A set of private companies also see land restoration as a potential business opportunity either with an interest to gain profits or a combination of profits and social/environmental impact.

However, much more needs to be attempted to see that the degradation of land is arrested and it is restored. Though certain policies work reasonably well in certain countries, the same may not be that effective in other contexts. Though there are global efforts like the Great Green Wall (GGW) initiative, the progress is not satisfactory. There are international declarations and allocations of money like that for addressing climate change and conserving biodiversity which can be used for land restoration, but these resources are yet to reach adequately the projects of land restoration, especially in contexts where these are most needed. There are signs of hope in private sector activities in restoration but most of these have not been able to scale up their operations.

Hence, the research behind of this book attempt to identify these persisting challenges which may work against land restoration at scale which is necessary to meet the global targets in this regard. These challenges and possible measures to overcome them are discussed in this concluding chapter.

V. Santhakumar et al., *Land Restoration: A Rational, Grounded and Pragmatic Approach*, https://doi.org/10.1007/978-3-032-26652-1_18

Poverty and Underdevelopment May Not Go Hand in Hand with Land Restoration

When people are extremely poor, the only resource that they may have access to is land and/or forests. They may use such land even if they don't have ownership of it. They may use it even if its fertility is very low. They will use all the labour that they can mobilise for that purpose. These can lead to land degradation. Poverty can also force them to discount future gains heavily and this can also encourage them to prefer short-term strategies to make a living (which may cause a higher level of degradation).

Moreover when people are in poverty and less-educated, the fertility rate of the population is likely to be higher. They need more hands for survival and they will not be able to generate much surplus for their old age life and children may serve as social security. When there is an increase in population, and when the land (or forests) that is accessible to them remain the same, there will be an intensive use of land which can aggravate its degradation. They may not be able to make investments for conservation of the land, say, through measures to control soil erosion. If they are practicing slash and burn cultivation, then they may be forced to use the same land within a relatively shorter period leading to a further decline in fertility and to severe degradation.

What about the intensive cultivation of large farms by better off sections of society? What about the urbanisation facilitated by industrial development? Won't these be degrading land? Yes these can also degrade land. However, there are a few differences. Usually, urbanisation leads to an intensive use of a smaller share of land in a country. Even if land is degraded in cities, the reduction in the pressure on land in the countryside can lead to the restoration of land at the aggregate level. Political economy is a lot more capable to regulate the actions of rich farmers who cultivate large stretches of land. Hence mandates such as the restoration of a part of private land, like those in Brazil, are possible in such a context. However, such regulations are not enforceable when a sizeable section of the population is poor and they use their land for their survival.

Need to Connect Land Restoration with the Goals of Economic and Human Development

We have seen that there is a need to address underdevelopment even for reducing land degradation and facilitating land restoration. Minimal social security for poorer people may help land restoration indirectly. If governments can ensure basic food for all people through a public distribution system, it may discourage people from using less-fertile land in a degrading manner.

An important investment for this purpose is the education of youngsters. Education can facilitate land restoration in multiple ways. The secondary education for girls may reduce the fertility rate and that can reduce population pressure on land in the long-run. When boys and girls get education, they may look for jobs in manufacturing and service sectors. If these jobs are not available in their own localities, they may migrate to cities. This can also facilitate international migration of workers if that is possible. Such a movement will reduce the pressure on land.

Though there will be a compulsion to spend more public resources in agriculture (due to the need to enhance domestic production of farm products and also due to the poverty and vulnerability of those who depend on agriculture), there will not be sustainable benefits when farm sizes are small (an indication of the dependence of higher population in that activity). Moreover, the lower income elasticity of agriculture would mean that the returns from it would be relatively lesser than that from industry and services. When people get educated, they will take up jobs in manufacturing and services, relieving pressure on agriculture and this can reduce the extensive agriculture. Though there is a possibility of the development of intensive agriculture, this may happen in relatively smaller share of land since lands which are less fertile may not be used for that purpose. Hence, the structural transformation of the economy that enables the movement of a higher share of population away from agriculture is desirable for land restoration. This is achieved as part of an inclusive economic and human development. There is a need to connect these development goals with those of land restoration.

Allocating Money and Actions at the Global Level Are Not Adequate; Need Attention to the Details and Capacity Building to See That It Reaches the Locations Where Restoration Is to Happen

Financing facilities for land restoration are increasingly available at international and national levels. However, the money that is reaching restoration efforts, especially in the less-developed and poorer countries, has not increased notably. We have looked at the peatland restoration in Indonesia as part of the work of the G20 Global Land Initiative. There were international and bilateral financing agreements aimed at the conservation of peatlands. However, the actual amount of resources which reach Indonesian agencies for actions on the ground is limited. There was a case when one such bilateral agreement was discontinued.

There could be multiple reasons for the ‘transmission loss’ between international declarations and the actual amount of money that reaches the projects of land restoration. First, international agreements may overestimate the difficulties in implementing projects by overlooking the context-specific institutional constraints. Secondly, the operational procedures for allocating money to specific projects may require greater clarity. There may be a need for changes in institutional regulations for the effective utilisation of the money. In certain other cases, the officials of less-developed countries may require greater capacity building to utilise these resources.

There is a need to address all these challenges to see that the resources which are available internationally and nationally reach the locations where these are needed, and the people on the ground who have to take actions.

Good to Have the Interest of Private Companies in Restoration; Need a Lot More Action and Support to Scale up Their Businesses

As part of the work of the G20 Global Land Initiative, we could see a number of private sector initiatives in land restoration. These include for-profit investments and those which combine profit motivation and a concern for natural environment. There are efforts to market products which are collected sustainably from forests or to grow plantations of indigenous species with plans for sustainable harvest of timber and other products. There are private investors who buy land and facilitate land restoration. The availability of finance through carbon and biodiversity credit, etc., should ideally enhance the interest of private players in land restoration.

These private initiatives are interesting and demonstrate that incentives at the level of individuals and firms need not be antithetical to land restoration. It can also facilitate innovation and the development of appropriate technology for the purpose. The development of technical production systems for the mass production of seedlings of indigenous tree varieties in Brazil by a company called Re-green is an example in this regard.

However, we note a set of limitations for these private sector initiatives. First, these initiatives altogether constitute only a very small part of actions which are needed for land restoration globally. Hence private initiatives as a whole are yet to be a major contributor to global restoration efforts. Most of these initiatives are small scale in nature. Many firms find it difficult to scale up their activities. The smaller size of the markets for products, lack of adequate financial investments, the presence of legal, institutional and policy hurdles, etc., work against the scaling up of these efforts. There is a potential to address some of these challenges through appropriate policy interventions globally and nationally.

However, the fundamental limitation of private efforts in land restoration seems to persist. This is due to the positive externality of land restoration. Those who carry out restoration may not be able to gain from this positive externality (which is gained by the society as whole). Though there are efforts to monetise these positive externalities (like the payment for ecosystem services) there are others which are yet to be or cannot be monetised. This may discourage for-profit private actors in restoration. Though many private initiatives have an element of altruism, it is also not adequate to provide land restoration to the level that is desired by the society.

Communities Have Mitigated Land Degradation in Certain Contexts in the Past; Need to Understand Their Potential and Challenges in Each Context

There were different views on the possible role of communities in the management of natural resources and such a management is important in the case of common property resources (CPR) like grasslands. The initial view was one of scepticism that is found in the concept of the 'Tragedy of Commons'. However, the empirical evidence collected by researchers (like Elinor Ostrom) indicated the possibility of sustainable management of common properties by communities. Economists have later brought out theoretical insights on situations where communities may manage CPR effectively. These include the presence of very-long mutually dependent interactions between members of the community and the absence of a very high discount rate.

There are many situations where such conditions prevail and one can see common properties including community lands managed reasonably well in such situations. The majority of indigenous communities in different parts of the world who still depend on natural ecosystems of their neighbourhood are of this kind. However, it would be naive to think that even these communities would remain in this equilibrium in future. We could see the weakening of community institutions and its impact on the sustainability of common lands in different parts of the world. These include the grasslands in mountainous areas of Central Asian countries like Kyrgyzstan.

All these communities which manage community lands are undergoing social transformation. Youngsters are getting educated and migrating to cities within or outside their countries. They send remittances and these have become a major source of income for those who remain in villages. There is an increasing dependence on goods sourced from distant markets for personal consumption. The mutual dependence among community members has declined. All these indicate the breaking down of one condition—very-long term interaction among members of the community—that has enabled the sustainable management of common property resources by communities.

The possible decline of the ability of communities to manage common lands sustainably may warrant newer institutional arrangements. In many cases, governments have stepped in, but the higher cost of public enforcement (and its associated inefficiencies) may create newer set of problems.

There Are Enabling and Disabling Policies of Governments; Need Context-Specific Policy Analysis

Policies for land restoration may work differently in different contexts. There can be incentives for individuals (farmers) for restoration. This is attempted in the US. However, these may work more effectively where land-to-people ratio is higher or where the dependence on land has come down as in the case of developed countries. It is incorrect to expect a similar impact of these policies in poorer countries where average farm size is very small. Similarly, there are reasonably effective legal measures which mandate land restoration in private lands in Brazil. Its success depends on the larger size of farms in the country. Even if such a law is adopted in countries like India where the average farm size is very small, it cannot be enforced easily.

Many less-developed countries have seen drastic deforestation as part of their economic growth. This was true even in the case of countries in Europe and Americas in the past when these were going through the process of economic development. It is also visible in the case of newly industrialised countries like China. However, the area of forests which was remaining in India when environmental concerns became prominent could be protected due to the state ownership of land (which was the outcome of an action of colonial rule), and also the control of an extensive government organisation (whose objective was the extraction of forest resources until then).

The way policies are enforced depends on social conditions, and these are also context-specific. If social incentives are against land restoration, then the cost of enforcement may go up even if a pro-restoration policy is adopted. This problem is true even if there are international efforts or resources for land restoration in a context. This is evident from the peatland restoration in Indonesia. Inadequate consideration of the incentives of local people to use land intensively is a reason for the failure of international initiatives of land restoration in different contexts.

The Impact of a Policy is Context-Specific Depending on the Presence/Absence of Other Enabling and Disabling Factors; Need a Systems Approach

Different socioeconomic factors may impact land restoration. However, the impact of one may depend on others, policies/institutions together and also the development context of the country. The contribution of any one factor, say private property rights, may be different in different contexts depending on the presence (or absence) of other enabling factors'.[1] This warrants a systems approach to understand the enablers of land restoration. This book attempts to outline such a systems approach towards restoration. Following this approach, there is a need to look at the parts and relationships which interact in the efforts towards restoring land individually and in an aggregate manner.

Such a systems approach is used to understand the situation in a set of countries in a couple of chapters of this book. There are interesting differences between these major countries. The higher per-capita land-size and the control of a major share of land by

---

[1] https://practiceconnect.azimpremjiuniversity.edu.in/envisioning-A-national-system-of-land-restoration/; https://practiceconnect.azimpremjiuniversity.edu.in/envisioning-a-national-system-of-land-restoration-part-2/.

the government are important enablers in Brazil whereas the cost of enforcing public ownership is a challenge there. The small size of the average land-holding and higher dependence on land (and a relatively lower human development) are the challenges in India whereas (historical) state ownership of forests is an enabling factor. Though forests were degraded drastically in China, industrial/economic development during the last four decades has enhanced the incentives and capabilities to adopt land restoration policies. All these show the importance of understanding how different socioeconomic factors interact in the context of each country.

Though There Are Developments in the Science of Land Restoration, There is a Need to Connect These Scientific Developments to Actions on the Ground

There is a need for research and science and technology development for land restoration. This may include techniques for the faster production of seedlings of indigenous (including wild) plants from different ecological settings. The negative impacts of exotic species also have to be monitored carefully. Interestingly, there are attempts to develop these technologies. For example, forestry scientists in collaboration with private investors have developed the technology to produce/multiply seedlings of indigenous trees in Brazil. Restoration efforts in other countries suffer from the lack of such technologies.

There are possibilities of using geospatial information and communication technology for land restoration. This may include that for identifying changes in the quality of land (whether as part of degradation or restoration) which can alert enforcement agencies and others, development of rating mechanisms which can facilitate law enforcement, the conversion of physical quantities into benefits (which are valuable in carbon or biodiversity markets), communication with wider stakeholders (like the information dissemination in share markets) and so on.

There may be an institutional reason for the slow propagation of such newer technologies. In the case of agriculture, farmers have an incentive to use new/better technology if it can lead to higher profits. (In the case of small farmers, governments may have the incentive to subsidise the availability of such technologies.) Similarly, industrial technology may witness a faster dissemination again due to the profit interest of firms. However, a major part of the benefits of land restoration can be public goods (and positive externality) which may be derived by society as a whole. Hence private incentives to adopt the technology are weaker in this regard. Hence, the key issue is the development of appropriate incentives for the use of science and technology for land restoration. This may require governmental efforts to encourage private actors to internalise the importance of land restoration. (Though governments have realised the importance of agricultural development and created incentives for the development and adoption of science and technology, land restoration is yet to be seen as a priority agenda). Given that land restoration can (needs to) be carried out through the actions of both the state and markets, the incentives for this purpose may have to develop in these two different institutional settings.

Ideal Land Restoration is Not Possible in All Contexts; Need to Focus on Marginal Benefits and Costs

Ecologists are concerned about the quality of land restoration. They may be concerned about the lack of enough efforts to protect biodiversity even when the land cover is restored. They may be concerned about increasing urbanisation even when rural lands encounter a decline of intensive use.

However, there is a need to realise that a perfect model of land restoration (even if there is one) is not going to happen in all parts of land. Land has multiple uses and some of these will work against such a perfect restoration. There will be cultivation, and the use of land for residence and industrial purposes, and also for commercial and entertainment uses. It may be possible to think about a situation where land is not degraded even when it is used for one or other purpose. That should be the strategy of land restoration in farm lands with the understanding that farming can cause certain impact on land. This may include the adoption of farming practices which may reduce soil erosion. Excessive use of fertiliser and pesticide can be controlled through regulation (and also through market signals. When people demand more of organic products as reflected in their price, farmers will have an incentive to reduce chemical fertilisers and pesticides.)

Certain restoration of land can happen with the conservation of land cover, though it may be with the help of only one or other species. Of course, such a strategy may not enhance biodiversity, but there is no increase in the marginal social benefit in pursuing biodiversity conservation in each and every parcel of land. Yes, there should be adequate in-situ conservation of biodiversity for all ecological types of land, but it should not be attempted in the whole stretch of land of each type. There may be cases where newer crops/plants may have to be introduced. There can be research and innovation in new crops so that their negative impacts on land can be minimised. Hence, what may be feasible (and hence desirable if we want to arrest land degradation in a timely manner) is to pursue different strategies of land restoration in different parcels of land, though all of these may not conform to an ideal model of restoration.

Need Education, Campaigns and Behavioural Change

There are global concerns about land degradation and international and national policy initiatives. However, the halting of degradation, and the restoring of land should happen on the ground in homesteads, farms and plantations.

The objective and subjective circumstances including poverty, attitudes and behaviours of individuals and families matter a lot for land. Hence there have to be an equal emphasis on changing micro-behaviour and actions. Local institutions including communities and local governments may have to be strengthened to enforce regulations and enhance incentives which encourage people to preserve land. There can be a higher level of diversity of these institutions which may warrant a diversity of approaches.

The concern for land should go hand in hand with the aspirations of ordinary people to have a better life. Advocacies to limit their consumption may not work. Unitary models of globalisation of production and consumption also may not be appropriate for the conservation of a variety of natural resources including the diverse varieties of land.

Healthy and productive land is to be seen as part of those attributes which can enable a good and happy life to people. Though the expansion of markets is important for economic development and such a development is important for land restoration too, markets which sustain the quality of land may be a lot more local. These indicate the importance of a diversity of consumption, production and exchange mechanisms that can sustain the diversity of life supporting mechanisms including land.

Need Effective Compensation Mechanisms

Though land restoration is beneficial to all, certain sections of people may have to incur a higher cost for this purpose. These may include those whose use of land has to be changed drastically and those who do not have other resources for their livelihood activities. Though this is known and discussed in the literature, there are not enough measures and mechanisms to compensate their losses. This can work against the effectiveness of land restoration projects in different parts of the world. Hence there should be a greater concern about the distributional aspects of land restoration in global and national programmes.

References

ABC. (n.d.). https://www.abc.net.au/reslib/200808/r282351_1199757.pdf

ABS. (2024a). National Land Account, Experimental Estimates, 2021 | Australian Bureau of Statistics. https://www.abs.gov.au/statistics/environment/environmental-management/national-land-account-experimental-estimates/latest-release#methodology

Abu Hammad, A., & Tumeizi, A. (2012). Land degradation: Socioeconomic and environmental causes and consequences in the eastern Mediterranean. *Land Degradation and Development, 23*, 216–226.

Ahlvik, L., & Sebastiaan, K. (2024). Distributional impacts of conservation on land prices: Evidence from natura 2000. https://doi.org/10.2139/ssrn.4938652

Albertus, M., Brambor, T., & Ceneviva, R. (2018). Land inequality and rural unrest: Theory and evidence from Brazil. *Journal of Conflict Resolution, 62*(3), 557–596.

Query ID="Q1" Text="" Alden Wily, L. (2017). Customary Tenure: Remaking Property for the 21st Century. In M. Graziadei & L. Smith (Eds.), *Comparative Property Law: Global Perspectives* (pp. 458–478) UK: Edgar Elgar.

Andrew, C., Jason, A., & David, C. (2017). Reflections on four decades of land restoration in Australia. *The Rangeland Journal, 39*, 405–416. https://doi.org/10.1071/RJ17056

Arneth, A., Olsson, L., Cowie, A., Erb, K. H., Hurlbert, M., Kurz, W. A., & Rounsevell, M. D. (2021). Restoring degraded lands. *Annual Review of Environment and Resources, 46*(1), 569–599.

Arrow, K. J. (1999a) Inter-generational equity and the rate of discount in long-term social investment. In Sertel, M. R. (Ed.) *Contemporary Economic Issues. International Economic Association Series* (pp. 89–102) London: Palgrave Macmillan

Asresehegn, T. G., Valencia, V., Schulz, S., Woldewahid, G., Gebrehawariat, G., & Schulte, R. P.O. (2025) Community responses to land degradation: Evidence from land restoration bright-spot communities in the Ethiopian highland, environmental development, 101422, ISSN 2211-4645. https://doi.org/10.1016/j.envdev.2025.101422

Australian Govt. (n.d.a). https://www.agriculture.gov.au/sites/default/files/documents/Indicator_1.1a-iv_Forest_area_change_over_time.pdf

Australian Govt. (n.d.c). 1.-DCCEEW-SOE_factsheet_Overview.pdf

Australian Govt. (n.d.b) https://www.agriculture.gov.au/sites/default/files/documents/Indicator_6_4a_Indigenous_forest_estate_2024.pdf

Bailkey, M., & Nasr, J. (2000). *From Brownfields to Greenfields: Producing Food in North American Cities*. Community Food Security News. Fall 1999/Winter 2000:6; Food for the Cities: Production systems (UPA). FAO.

Balmford, A., Bruner, A., Cooper, P., Costanza, R., Farber, S., Green, R. E., & Turner, R. K. (2002). Economic reasons for conserving wild nature. *Science, 297*(5583), 950–953

V. Santhakumar et al., *Land Restoration: A Rational, Grounded and Pragmatic Approach*, https://doi.org/10.1007/978-3-032-26652-1

Barbier, E. B. (1997). The economic determinants of land degradation in developing countries. *Philosophical Transactions of the Royal Society, London, B352*, 891–899. https://doi.org/10.1098/rstb.1997.0068

Barbier, E. B., & Di Falco, S. (2021). Rural populations, land degradation, and living standards in developing countries. *Review of Environmental Economics and Policy, 15*(1), 115–133.

Barrett, et al. (2002). Non-farm income diversification and household livelihood strategies in rural Africa: Concepts, dynamics, and policy implications. *Food Policy, 26*(4), 315–331.

Barrow, C. J. (1991). Land degradation: Development and breakdown of terrestrial environments. CombridgeUniv Press, Cambridge. p. 295.

Benartzi, S., & Thaler, R. H. (1995). Myopic loss aversion and the equity premium puzzle. *Quarterly Journal of Economics, 110*, 73–92.

Benartzi, S., & Thaler, R. H. (2004). Save more tomorrow: Using behavioral economics to increase employee saving. *Journal of Political Economy, 112*, 164–187.

Bergstrom, T. C. (1973). Private demand for public goods. *American Economic Review, 63*(3), 280–296.

Bourdieu, P. (1984). *Distinction: A social critique of the judgement of taste*. Harvard University Press.

Chasek, P., Akhtar-Schuster, M., Orr, B. J., Luise, A., Rakoto Ratsimba, H., & Safriel, U. (2019). Land degradation neutrality: The science-policy interface from the UNCCD to national implementation. *Environmental Science & Policy, 92*, 182–190.https://doi.org/10.1016/j.envsci.2018.11.017

Chazdon, R. L., Wilson, S. J., Brondizio, E., Guariguata, M. R., & Herbohn, J. (2021). Key challenges for governing forest and landscape restoration across different contexts. *Land Use Policy, 104*, Article 104854.

Collard, S. J., O'Connor, P. J., Prowse, T. A. A., Gregg, D., & Bond, A. J. (2019). Objectives versus realities: spatial, temporal, financial and social deficiencies in Australia's public revegetation investment model. *Ecological Management & Restoration, 21*(1), 35–41. https://doi.org/10.1111/emr.12398

Cook-Patton, S. C., Drever, C. R., Griscom, B. W., et al. (2021). Protect, manage and then restore lands for climate mitigation. *Nature Clinical Practice Endocrinology & Metabolism, 11*, 1027–1034. https://doi.org/10.1038/s41558-021-01198-0

Cullen, J. (2018). *After HLEG: The Prospects for Sustainable Finance in Europe*. The Cambridge Yearbook of European, Legal Studies.

Dai, H., Zhu, Z., Trachung, B., Golog, D., Riley, M., Lü, Z., & Li, L. (2024). Communities in ecosystem restoration: The role of inclusive values and local elites' narrative innovations. *People and Nature, 6*, 1655–1667. https://doi.org/10.1002/pan3.10675

DCCEEW. (n.d.a) Outback Australia—the rangelands—DCCEEW. https://www.dcceew.gov.au/environment/land/rangelands

Designing financial instruments for land-based ecological restoration: A review and future research Agenda. *Cleaner Production Letters, 8*. https://doi.org/10.1016/j.clpl.2024.100089

Devi, S., Goswami, S., Bisaria, J., & Sinha, B. (2024). Approaches for land restoration: An alternative strategy for climate change mitigation. In: Singh, H. (Eds.), *Forests and Climate Change*. Singapore: Springer. https://doi.org/10.1007/978-981-97-3905-9_26

Ding, H., et al. (2017). *Roots of prosperity: The economics and finance of restoring land*. World Resources Institute.

Dohong, A., Abdul Aziz, A. & Dargusch, P. (2018). A review of techniques for effective tropical peatland restoration. *Wetlands, 38*, 275–292. https://doi.org/10.1007/s13157-018-1017-6

Duchelle, A. E., Cromberg, M., Gebara, M. F., Guerra, R., Melo, T., Larson, A., & Sunderlin, W. D. (2014). Linking forest tenure reform, environmental compliance, and incentives: lessons from REDD+ initiatives in the Brazilian Amazon. *World Development, 55*, 53–67.

Edwards, D. P., & Cerullo, G. R. (2024). Biodiversity is central for restoration. *Current Biology, 34*(9), R371–R379.

Egidi, G., Zambon, I., Tombolin, I., Salvati, L., Cividino, S., Seifollahi-Aghmiuni, S., & Kalantari, Z. (2020). Unraveling latest aspects of urban expansion: Desertification risk reveals more. *International Journal of Environmental Research and Public Health (MDPI), 17*, 4001.

Elobeid, A., Carriquiry, M., Dumortier, J., Swenson, D., & J Hayes, D. (2021). China-US trade dispute and its impact on global agricultural markets, the US economy, and greenhouse gas emissions. *Journal of Agricultural Economics, 72*(3), 647–672.

Farhadinia, M. S., Waldron, A., Kaszta, Ż., Eid, E., Hughes, A., Ambarlı, H., & Macdonald, D. W. (2022). Current trends suggest most Asian countries are unlikely to meet future biodiversity targets on protected areas. *Communications Biology, 5*(1), 1221.

Fehr, E., & Henrich, J. (2003). Is strong reciprocity a maladaptation? On the evolutionary foundations of human altruism. In P. Hammerstein (Ed.), *Genetic and cultural evolution of cooperation* (pp. 55–82). MIT Press.

Ferraro, P. J., & Kiss, A. (2002). Direct payments to conserve biodiversity. *Science* (New York, N.Y.). 298, 1718–9. https://doi.org/10.1126/science.1078104.

Fitz, D. (2018). Evaluating the impact of market-assisted land reform in Brazil. *World Development, 103*, 255–267.

Fox, R. C., & Rowntree, K. M. (2001). Redistribution, restitution and reform: Prospects for the land in the Eastern Cape Province, South Africa. In Conacher, A. J. (Ed.), Land Degradation. *The GeoJournal Library* (vol. 58, pp. 167–186). Springer.

François, D., Medous, L., & Etrillard, C. (2024). Use of linear transportation infrastructure rights-of-way as an ecological shelter: National asset estimate and stakeholder involvement. *Nature Conservation, 57*, 17–40.

Gambella, F., Quaranta, G., Morrow, N., Vcelakova, R., Salvati, L., Gimenez Morera, A., & Rodrigo-Comino, J. (2021). Soil Degradation and Socioeconomic Systems' Complexity: Uncovering the Latent Nexus. *Land, 10*(1), 30. https://doi.org/10.3390/land10010030

Gerber, N., Nkonya, E., & von Braun, J. (2014). Land degradation, poverty and marginality. In: von Braun, J., Gatzweiler, F. (eds) *Marginality*. Dordrecht: Springer. https://doi.org/10.1007/978-94-007-7061-4_12

Gilbert, N. (2012). Organic farming is rarely enough. *Nature*. https://doi.org/10.1038/nature.2012.10519

GoSA. (n.d.). Aboriginal firestick farming. https://www.education.sa.gov.au/our-learning-sa/learning-home-activities/science/aboriginal-firestick-farming#:~:text=Fire%252C%2520the%2520result%2520of%2520a,practice%2520known%2520as%2520firestick%2520farming

Griskevicius, V., & Tybur, J. M. (2010). Going green to be seen: Status, reputation, and conspicuous conservation. *Journal of Personality and Social Psychology, 98*(3), 392–404.

Gunter, U., & Smeral, E. (2016). The decline of tourism income elasticities in a global context. *Tourism Economics., 22*, 466–483. https://doi.org/10.5367/te.2014.0431

Guth, W., Schmittberger, R., & Schwartz, B. (1982). An Experimental Analysis of Ultimatum Bargaining. *Journal of Economic Behaviour and Organization, 3*(4), 367–388.

Harada, K., Habib, M., Sakata, Y., Maryudi, A. (2022). The role of NGOs in recognition and sustainable maintenance of customary forests within indigenous communities: The case of Kerinci, Indonesia. *Land Use Policy, 113*, 105865, ISSN 0264-8377. https://doi.org/10.1016/j.landusepol.2021.105865. https://www.sciencedirect.com/science/article/pii/S0264837721005883

Hardin, G. (1968, December) The tragedy of the commons, science, *New Series, 162*(3859), 1243–1248.

Haushofer, J., & Fehr, E. (2014). On the psychology of poverty. *Science, 344*, 862–867.https://doi.org/10.1126/science.1232491

Hausman, J. A. (1979). Individual discount rates and the purchase and utilization of energy-using durables. *The Bell Journal of Economics, 10*(1), 33–54.

Herrick, J.E., Abrahamse, T., Abhilash, P.C., Ali, S.H., Alvarez-Torres, P., Barau, A.S., Branquinho, C., Chhatre, A., Chotte, J.L., Cowie, A.L., Davis, K.F., Edrisi, S.A., Fennessy, M.S., Fletcher, S., Flores-Díaz, A.C., Franco, I.B., Ganguli, A.C., Ifejika Speranza, C., Kamar, M.J., Kaudia, A.A., Kimiti, D.W., Luz, A.C., Matos, P., Metternicht, G., Neff, J., Nunes, A., Olaniyi, A.O.,

Pinho, P., Primmer, E., Quandt, A., Sarkar, P., Scherr, S.J., Singh, A., Sudoi, V., von Maltitz, G.P., Wertz, L., Zeleke, G. A think piece of the International Resource Panel. United Nations Environment Programme, Nairobi, Kenya

Hirsh, J. B. (2010). Personality and environmental concern. *Journal of Environmental Psychology, 30*(2), 245–248.

Hoffman, M. T., & Todd, S. (2000). A national review of land degradation in South Africa: The influence of biophysical and socio-economic factors. *Journal of Southern African Studies, 26*(4), 743–758.

Inglehart, R. (1977). *The Silent Revolution: Changing Values and Political Styles Among Western Publics*. Princeton University Press.

International Resource Panel. (2019). Land Restoration for Achieving the Sustainable Development Goals: An International Resource Panel Think Piece.

Istanbuly, M.N., Krása, J., Jabbarian, A. B. (2022, February) How socio-economic drivers explain landscape soil erosion regulation services in polish catchments. *International Journal of Environmental Research Public Health, 19*(4):2372. https://doi.org/10.3390/ijerph19042372. PMID: 35206558; PMCID: PMC8872238.

John, G., Grigoriadis, V., Flannery, D., Knapp, E., Gold, E., Hutchinson, G., Frewer, L. J., Brereton, P., Byrne, K. A. (2025).

Jordan, W. R., Peters, R. L., & Allen, E. B. (1988). Ecological restoration as a strategy for conserving biological diversity. *Environmental Management, 12*, 55–72.

Kahneman, D., Knetsch, J. L., & Thaler, R. (1986). Fairness as a constraint on profit-seeking: Entitlements in the market. *American Economic Review, 76*(4), 728–741.

Kiely, L., Spracklen, D. V., Arnold, S. R., Papargyropoulou, E., Conibear, L., Wiedinmyer, C., Adrianto, H. A. (2021). Assessing costs of Indonesian fires and the benefits of restoring peatland. *Nature Communications, 12*, 7044.

Kotchen, M. (2012). Public Goods. In J. Whitehead, & T. Haab (Eds.), *Environmental and Natural Resource Economics: An Encyclopaedia*, Santa Barbara, CA: ABC-CLIO, Inc

Kugler, J. C., Smith, T., Connolly, Y.-J., & Son (Eds.), *Decision Modeling and Behavior in Uncertain and Complex Environments* (pp. 57–76). Berlin: Springer

Naveed, et al. (2022). What have we learned from Environmental Kuznets Curve hypothesis? A citation-based systematic literature review and content analysis. *Energy Strategy Reviews, 44*

Le, T. -A. T., Vodden, K., Wu, J., Bullock, R., Sabau, G. (2024). Payments for ecosystem services programs: A global review of contributions towards sustainability. *Heliyon, 10*(1), e22361, ISSN 2405-8440, https://doi.org/10.1016/j.heliyon.2023.e22361. (https://www.sciencedirect.com/science/article/pii/S2405844023095695)

Li, B. V., & Pimm, S. L. (2020). How China expanded its protected areas to conserve biodiversity. *Current Biology, 30*(22), R1334–R1340.

Löfqvist, S., et al. (2023a). How social considerations improve the equity and effectiveness of ecosystem restoration. *BioScience, 73*, 134–148.

Löfqvist, S., Garrett, R. D., & Ghazoul, J. (2023b). Incentives and barriers to private finance for forest and landscape restoration. *Nature Ecology & Evolution, 7*, 707–715. https://doi.org/10.1038/s41559-023-02037-5

Lundvall, B. A. (1992). *National systems of innovation: Towards a theory of innovation and interactive learning* (Vol. 242). Pinter.

Ma, Z., Xia, C., & Cao, S. (2020). Cost-benefit analysis of China's natural forest conservation program. *Journal of Nature Conservation*, Article 125818. https://doi.org/10.1016/j.jnc.2020.125818

Mani, S., Osborne, C. P., & Cleaver, F. (2021). Land degradation in South Africa: Justice and climate change in tension. *People and Nature, 3*(5), 978–989.

Mani, S., Mohanakumar, S. M., Santhakumar, V., Abhilash, T. (2020). Conservation of agrobiodiversity: Assessing the policies and institutions in Kerala. In Mani, S. (Ed.), *Kerala and the World Economy, Centre for Development Studies, Trivandrum*

Mappin, B., Ward, A., Hughes, L., Watson, J., Cosier, P., & Possingham, H. P. (2021). The costs and benefits of restoring a continent's terrestrial ecosystems. *Journal of Applied Ecology, 59*(2), 408–419. https://doi.org/10.1111/1365-2664.14008

Margulies, J. D., & Karanth, K. K. (2018). The production of human-wildlife conflict: A political animal geography of encounter. *Geoforum, 95*, 153–164.

Maskin, E. (1986). The folk theorem in repeated games with discounting or with incomplete information. *Econometrica, 54*(3), 533–554.

McFarland, W., Whitley, S., & Kissinger, G. (2015). *"Subsidies to Key Commodities Driving Forest Loss: Implications for Private Climate Finance." Working Paper*. London: Overseas Development Institute

Meadows, M. E., & Hoffman, M. T. (2002). The nature, extent and causes of land degradation in South Africa: Legacy of the past, lessons for the future? *Area, 34*(4), 428–437.

Mirzabaev, A., & Wuepper, D. (2023). Economics of ecosystem restoration. *Annual Review of Resource Economics, 15*(1), 329–350.

Moon, S., Arul, M., Himanshu, M., & Moon, K. (2016). Cultural and economic impacts on global cultural products: Evidence from US movies. *Journal of International Marketing, 24*. https://doi.org/10.1509/jim.15.0080

Namubiru-Mwaura, E. (2021). *Gender and Land Restoration*. UNCCD Global Land Outlook Working Paper. Bonn.

Nepal, A, K., Nepal, M., & Bluffstone, R. (2022) *Labour Outmigration, Farmland Fallowing, Livelihood Diversification and Technology Adoption in Nepal*. https://doi.org/10.1111/ilr.12375

Oliveira, P. T. S., de Faria Godoi, R., Colman, C. B., Motta, J. S., Sone, J. S., & Almagro, A. (2022). Agricultural land degradation in Brazil. In *Impact of Agriculture on Soil Degradation I: Perspectives from Africa, Asia, America and Oceania* (pp. 97–127). Cham: Springer International Publishing.

Pfeifer, M. et al. (2023) A systems approach framework for evaluating tree restoration interventions for social and ecological outcomes in rural tropical landscapes. *Philosophical Transactions of the Royal Society B Biological Science, 378*, 20210111.

Podestá, G., Weber, E. U., Laciana, C., Bert, F., & Letson, D. (2008). Agricultural decision-making in the Argentine Pampas: Modeling the interaction between uncertain and complex environments and heterogeneous and complex decision-makers.

Pollitt, M. G., & Shaorshadze, I. (2011). *The Role of Behavioural Economics in Energy and Climate Policy*, Cambridge Working Paper in Economics 1165, Cambridge University.

QLD. (n.d.b). The Land Restoration Fund | Environment, land and water | Queensland Government

Rapiya, M., Truter, W., & Ramoelo, A. (2024). The integration of land restoration and biodiversity conservation practices in sustainable food systems of Africa: A systematic review. *Sustainability, 16*(20), 8951.

Reserve Bank of Australia. (2025). Composition of the Australian Economy Snapshot | Education | RBA. https://www.rba.gov.au/education/resources/snapshots/economy-composition-snapshot/#

Rodiger, M., & Hamm, U. (2015) How are organic food prices affecting consumer behaviour? A review. *Food Quality & Preference, 43*

Rubaltelli, E., Pasini, G., Rumiati, R., Olsen, R. A., & Slovic, P. (2010). The influence of affective reactions on investment decisions. *Journal of Behavioural Finance, 11*, 168–176.

Ryan, R. M., & Doci, E. L. (2000). Intrinsic and extrinsic motivations: Classic definitions and new directions. *Contemporary Educational Psychology, 25*, 54–67.

Salvati, L., & Zitti, M. (2007). Territorial disparities, natural resource distribution, and land degradation: A case study in southern Europe. *GeoJournal, 70*, 185–194.

Santhakumar, V. (2017). The state of India's environment: Reflecting its underdevelopment. *Sustainability in Debate, 8*(3), 75–83.

Santhakumar, V. (2009). Entry fees as an instrument for environmental management study of a wildlife park in Kerala, India. *Tourism Economics, 15*(2).

Sawosri, A. W., & Mußhoff, O. (2020). Risk and time preferences of farmers in India and Indonesia. EFForTS Discussion Paper Series 32, University of Goettingen, Collaborative Research Centre 990. EFForTS, Ecological and Socioeconomic Functions of Tropical Lowland Rainforest Transformation Systems (Sumatra, Indonesia).

Schubert, C. (2017). Green nudges: Do they work? Are they ethical? *Ecological Economics, 132*, 329–342.

Seifollahi-Aghmiuni, S., Kalantari, Z., Egidi, G., et al. (2022). Urbanisation-driven land degradation and socioeconomic challenges in Peri-urban areas: Insights from Southern Europe. *Ambio, 51*, 1446–1458. https://doi.org/10.1007/s13280-022-01701-7

Sharafatmandrad, M., & Mashizi, A. K. (2021). Linking ecosystem services to social well-being: an approach to assess land degradation. *Frontiers in Ecology and Evolution, 9*. https://doi.org/10.3389/fevo.2021.654560

Shi, M. J., & Chen, K. (2004). Land degradation, government subsidy, and smallholders' conservation decision: the case of the loess plateau in China. *Journal of Zheijang University-Science A, 5*, 1533–1542. https://doi.org/10.1631/jzus.2004.1533

Sigman, E., & Elias, M. (2021). Three approaches to restoration and their implications for social inclusion. uwpress. org. *Ecological Restoration, 39*, 1–2

Simonson, W. D., Miller, E., Jones, A., García-Rangel, S., Thornton, H., & McOwen, C. (2021). Enhancing climate change resilience of ecological restoration—A framework for action. *Perspectives in Ecology and Conservation, 19*(3), 300–310.

Sklenicka, P., Zohar, J., Molnarova, K. J., Vlasak, J., Kottova, B., Petrzwlka, P., Gebhart, M., & Walmsley, A. (2020). Trends of soil degradation: Does the socio-economic status of land owners and land users matter? *Land Use Policy, 95*, Article 103992. https://doi.org/10.1098/rstb.1997.0068

Smit, J., Ratta, A., Nasr, J. (1996). Urban agriculture: food, jobs, and sustainable cities. New York, N.Y.: United Nations Development Programme. ISBN 9789211260472. OCLC 34575217.

SoE. (2021). Population | Australia state of the environment 2021. https://soe.dcceew.gov.au/coasts/pressures/population

Zu Ermgassen, S. O. S. E., Löfqvist, S. (2024). Financing ecosystem restoration. *Current Biology, 34*(9), R412-R417, ISSN 0960-9822, https://doi.org/10.1016/j.cub.2024.02.031. https://www.sciencedirect.com/science/article/pii/S0960982224001738

Strassburg, B. B. N., Iribarrem, A., Beyer, H. L. et al. (2020). Global priority areas for ecosystem restoration. *Nature, 586*, 724–729. https://doi.org/10.1038/s41586-020-2784-9. https://sdg-action.org/restoring-land-to-combat-climate-change-and-degradation/#:~:text=According%20to%20Strassburg%20et%20al,conservation%20and%20climate%20mitigation%20agendas

Sze, J. S., Carrasco, L. R., Childs, D., & Edwards, D. P. (2022). Reduced deforestation and degradation in Indigenous Lands pan-tropically. *Nat. Sustain., 5*, 123–130. https://doi.org/10.1038/s41893-021-00815-2

Tanaka, T., Camerer, C. F., & Nguyen, Q. (2016). Risk and time preferences: Linking experimental and household survey data from Vietnam. In S. Ikeda, H. K. Kato, F. Ohtake, & Y. Tsutsui (Eds.), *Behavioral Economics of Preferences, Choices, and Happiness* (pp. 3–25). Springer.

UNCCD United Nations Convention to Combat Desertification. (2022). The Global Land Outlook, Second edn. UNCCD, Bonn.

UNCCD. (2022a). UNCCD Australia report. https://data.unccd.int/country-overview?country=AUS&layer=VOLUNTARY_TARGETS

UNCCD. (2022b). Microsoft Word—Glossary_2021_EN_2021-10-28_noSources.docx. https://www.unccd.int/sites/default/files/inline-files/Glossary_UNCCD-2022-reporting-process_EN.pdf

UNEP (United Nations Environment Programme). (2023). State of Finance for Nature: The Big Nature Turnaround—Repurposing $7 trillion to combat nature loss. Nairobi. https://doi.org/10.59117/20.500.11822/44278

United Nations Convention to Combat Desertification (2017). The Global Land Outlook, first edition. Bonn, Germany.

Verdone, M., & Seidl, A. (2017). Time, space, place, and the Bonn challenge global forest restoration target. *Restoration Ecology, 25*(6), 903–911.

Verhoeven, D., Berkhout, E., Sewell, A., & Esch, S. (2024). The global cost of international commitments on land restoration. *Land Degradation & Development., 35*, 4864–4874. https://doi.org/10.1002/ldr.5263

von Braun, J., Gerber, N., Mirzabaev, A., & Nkonya, E. (2013). The economics of land degradation (No. 109). ZEF Working paper series

Vu, Q. M., Lea, Q. B., Frossardd, E., & Vlek, P. L. G. (2014). Socio-economic and biophysical determinants of land degradation in Vietnam: An integrated causal analysis at the national level. *Land Use Policy, 36*, 605–617.

Warren, M., Hergoualc'h, K., Kauffman, J. B., Murdiyarso, D., Kolka, R. (2017, December). An appraisal of Indonesia's immense peat carbon stock using national peatland maps: uncertainties and potential losses from conversion. *Carbon Balance Manage, 12*(1):12. https://doi.org/10.1186/s13021-017-0080-2. Epub 2017 May 19. PMID: 28527145; PMCID: PMC5438333.

Weber, E. U., Johnson, E. J., Milch, K., Chang, H., Brodscholl, J., & Goldstein, D. (2007). Asymmetric discounting in intertemporal choice: A query theory account. *Psychological Science, 18*, 516–523.

Weber, E. U., & Johnson, E. J. (2012). *Psychology and Behavioral Economics Lessons for the Design of a Green Growth Strategy*, Working Paper No. 6240, The World Bank, Washington, D.C.

World Development Report. (2015). Mind, Society, and Behavior, Washington, D. C.: The World Bank Group.

World Resources Institute. 2016. Can Plantations Help Restore Degraded and Deforested Land? https://www.wri.org/insights/can-plantations-help-restore-degraded-and-deforested-land

Wuepper, D., Borrelli, P., Panagos, P., Lauber, T., Crowther, T., Thomas, A., & Robinson, D. A. (2021). A 'debt'based approach to land degradation as an indicator of global change. *Global Change Biology, 27*(21), 5407–5410.

Wuepper, D., Crowther, T., Lauber, T., Routh, D. L., Clech, S., Garrett, R. D., & Borner, J. (2024a). Public policies and global forest conservation: Empirical evidence from national borders. *Global Environmental Change, 84*, Article 102770.

Wuepper, D., Wiebecke, I., Meier, L., Vogelsanger, S., Bramato, S., Fürholz, A., & Finger, R. (2024). Agri-environmental policies from 1960 to 2022. *Nature food, 5*(4), 323–331.

Yuwati, T. W., Rachmanadi, D., Pratiwi, Turjaman, M., Indrajaya, Y., Nugroho, H. Y. S. H., & Mendham, D. (2021). Restoration of degraded tropical peatland in Indonesia: A review. *Land, 10*(11), 1170.

Zashimuddin, M. (2004). Community forestry for poverty reduction in Bangladesh. Forests for poverty reduction: Can community forestry make money? 81–94.

Zhai, D., Xu, J., Dai, Z., & Schmidt-Vogt, D. (2017). Lost in transition: Forest transition and natural forest loss in tropical China. *Plant Diversity, 39*(3), 149–153.

Zhang, K. H., & Shunfeng, S. O. N. G. (2003). Rural–urban migration and urbanization in China: Evidence from time-series and cross-section analyses. *China Economic Review, 14*(4), 386–400.

The manufacturer's authorised representative in the EU is Springer Nature Customer Service Centre GmbH, Europaplatz 3, 69115 Heidelberg, Germany. If you have any concerns regarding our products, please contact ProductSafety@springernature.com

Printed and bound by CPI Group (UK) Ltd, Croydon, CR0 4YY
07/10/2026
02229366-0003